Missouri Total Eclipse Guide

Official Commemorative 2024 Keepsake Guidebook

2024 Total Eclipse State Guide Series

Aaron Linsdau

Sastrugi Press

Jackson Hole

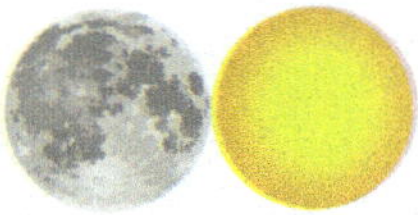

Sastrugi Press / Published by arrangement with the author

Missouri Total Eclipse Guide: Official Commemorative 2024 Keepsake Guidebook

The author has made every effort to accurately describe the locations contained in this work. Travel to some locations in this book is hazardous. The publisher has no control over and does not assume any responsibility for author or third-party websites or their content describing these locations, how to travel there, nor how to do it safely. Refer to local regulations and laws.

Any person exploring these locations is personally responsible for checking local conditions prior to departure. You are responsible for your own actions and decisions. The information contained in this work is based solely on the author's research at the time of publication and may not be accurate in the future. Neither the publisher nor the author assumes any liability for anyone climbing, exploring, visiting, or traveling to the locations described in this work. Climbing is dangerous by its nature. Any person engaging in mountain climbing is responsible for learning the proper techniques. The reader assumes all risks and accepts full responsibility for injuries, including death.

Sastrugi Press
PO Box 1297, Jackson, WY 83001, United States
www.sastrugipress.com
Quantity sales: Special discounts are available on quantity purchases by corporations, associations, and others. For details, contact the publisher at the address above.

Library of Congress Catalog-in-Publication Data
Library of Congress Control Number: 2019949785
Linsdau, Aaron
Missouri Total Eclipse Guide / Aaron Linsdau-1st United States edition
p. cm.
1. Nature 2. Astronomy 3. Travel 4. Photography
Summary: Learn everything you need to know about viewing, experiencing, and photographing the total eclipse in Missouri on April 8, 2024.

ISBN-13: 978-1-64922-317-3 (paperback)

508.4—dc23

All photography, maps and artwork by the author, except as noted.

10 9 8 7 6 5

Contents

Introduction

Thank you for purchasing this book. It has everything you need to know about the total eclipse in Missouri on April 8, 2024.

A total eclipse passing through the United States is a rare event. The last US total eclipse was in 2017. It traveled from Oregon to South Carolina. The last American total eclipse prior to that was in 1979!

The next total eclipse over the US will not be until April 8, 2024. It will pass over Texas, the Midwest, and on to Maine. After that, the next coast-to-coast total eclipse will be in 2045.

It's imperative to make travel plans early. You will be amazed at the number of people swarming to the total eclipse path. Some might say watching a partial versus a total eclipse is a similar experience. It's not.

This book is written for Missouri visitors and anyone else viewing the eclipse. You will find general planning, viewing, and photography information inside. Should you travel to the eclipse path in Missouri in April, be prepared for an epic trip. The estimates based on the 2017 eclipse suggest that millions will converge on Missouri.

Some hotels in the communities and cities along the path of totality in Missouri have already been contacted by people to make reservations. Finding lodging along the eclipse path may be a major challenge.

Resources will be stretched far beyond the normal limits. Think gas lines from the late 1970s. It may be likely that traffic along highways will come to a complete standstill during this event. Be prepared with backup supplies.

Many smaller Missouri towns are far from any major city. Missouri country roads can be slow. Please obey posted speed limits for the safety of everyone. Be cautious about believing a map application's estimate of travel time in Missouri.

People in communities along the path of the total eclipse may rent out properties for this event. With this major celestial spectacle in the spring of 2024, be assured that Missouri "hasn't seen anything yet."

Is this to say to avoid Missouri or other areas during the eclipse? Not at all! This guidebook provides ideas for interesting, alternative, and memorable locations to see the eclipse. It will be too late to rush to a better spot once the eclipse begins. Law enforcement will be out to help drivers reconsider speeding.

Please be patient and careful. There will be a large rush of people from all over the world, converging on Missouri to enjoy the total eclipse. Be mindful of other drivers on eclipse weekend, as they may not be familiar with Missouri roads.

You should feel compelled to play hooky on April 8. Ask for the day off. Take your kids out of school. They'll be adults before the next chance to see a total eclipse over America. Create family memories that will last a lifetime. Sastrugi Press does not normally advocate skipping school or work. Make an exception because this is too big an event to miss.

Wherever you plan to be along the total eclipse path, leave early and remember your eclipse glasses. People from all around the planet will converge on Missouri. Be good to your fellow humans and be safe. We all want to enjoy this spectacular show.

Visit www.sastrugipress.com/eclipse for the latest updates for this state eclipse book series.

Author Information

Polar explorer and motivational speaker Aaron Linsdau's first book, *Antarctic Tears*, is an emotional journey into the heart of Antarctica. He ate two sticks of butter every day to survive. Aaron coughed up blood early in the expedition and struggled with equipment failures. Despite the endless difficulties, he set a world record for surviving the longest solo expedition to the South Pole.

Aaron teaches how to build resilience to overcome adversity by managing attitude. He shares his techniques for overcoming adrenaline burnout and constant overload. He inspires audiences to face their challenges with a new perspective. As a motivational speaker, Aaron talks about courage, resilience, attitude, safety, and risk. He hopes that you will be inspired and have an enjoyable time watching the total eclipse in Missouri.

Visit his websites at www.aaronlinsdau.com or www.ncexped.com.

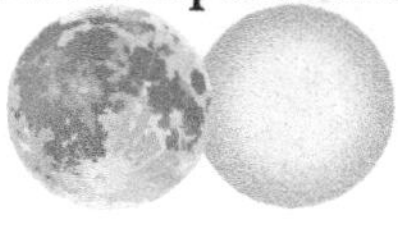

All About Missouri

Overview of Missouri

Missouri, the Show-Me State, welcomes visitors as if greeting an old friend. Whether you are looking for an outdoor vacation or a culturally educational experience, this diverse state will be just what the doctor ordered. Museums are as widespread and varied, appealing to the classical art enthusiast or a more free-spirited individual.

Located in southern Missouri, Branson opens up the possibilities of history with a visit to the Titanic Museum. For those who prefer a peek into a more bizarre and uncanny experience, a stop at Ripley's Believe It or Not Museum is a spectacular experience for all ages. The glitzy world of Branson offers attractions to please the senses.

Adventures in southern Missouri continue with Silver Dollar City attractions designed for the entertainment of visitors of all ages and diversity. The city's artisans portray the culture of the Ozarks, treating people to the wonder of glassblowing, knife making, weaving, candle making, and a multitude of crafting well known to the Ozark Mountain region.

Musical theater offerings line the brightly lit streets of Branson, showcasing a cultural extravaganza. Branson's theater patrons can experience the world-famous Baldknobbers that will entreat folks with a mixture of extraordinary country music and a dash of slapstick comedy that will leave an ache in your funny bone. Country music stands as the backdrop of this bright shining small town with big-city flair.

Situated in the middle of the state, Lake of the Ozarks brings both the fishing world as well as boating enthusiasts a backdrop like no other in the country. The crystal-clear water is a perfect location for enjoying a variety of water sports such as boating, waterskiing, jet skiing, and fly boarding. Whether visitors are experienced experts or new to the world of water sports, Missouri's Lake of the Ozarks is a perfect playground.

Northern Missouri offers the background atmosphere into the world of legendary Mark Twain. Stop in for a visit to picturesque Hannibal for a look into the infamous author's hometown, with a trip to his boyhood home and museum. Attractions around the life of one of America's favorite literary influences are brought to the visitor in a Riverboat cruise on the mighty Mississippi, letting the bystander take a peek into the life of Tom Sawyer. But Hannibal presents a multitude of allure with shopping, art museums, theater events or just a simple stroll among the monuments.

Glitzy casinos situated with the backdrop of the Gateway Arch provide visitors to St. Louis excitement from a sparkle cast by one of the most famous landmarks. The Arch is symbolic of the gateway to the west, allowing a view into the world of the past while St. Louis spread options such as the St. Louis Zoo. Situated with the Zoo are the National Historic Botanical Garden, one of the oldest in the country, spreading beauty for the guests.

There are many state parks and caves scattered across the state, allowing visitors the joy of exploration. Elephant Rock States Park lays holds a selection of magnificent rock formations that continue to delight the world. It provides spelunkers with underground adventures such as Marvel Cave, situated amid the world of Silver Dollar City. It has guided tours through the splendor below the surface. Explore Onondaga Cave in the countryside of rural Leasburg near St. Louis. Or sit back and relax with a jeep tour through Green County, Missouri's ever-famous Fantastic Caverns. Meramec Caverns, Talking Rocks, Bridal Cave, and Smallin Civil War Cave are just a portion of the caving adventure Missouri holds within its borders.

Theme parks, museums, national parks, and shopping are just some of the splendid attractions within this wondrous destination, while a varied wildlife population brings another world to life. All ages may take in the spectacular spring flowers for a more relaxing journey across Missouri or sit back on the historic Route 66. Tour the state with its flair for familiarity that the locals portray with their easygoing friendly charm, making Missouri a dream destination for everyone.

Hotels and Motels During the Eclipse

Once excitement of the total eclipse over Missouri spreads, rooms will become scarce. Many hotels in towns along the path of totality in western states sold out for a year or more during the 2017 total eclipse. Missouri is not alone in this challenge. Hotels all along the path of totality will sell out in anticipation of the 2024 total eclipse.

What does this mean for eclipse visitors? Lodging and room rentals in eclipse towns will be at a massive premium. Does that mean all hope is lost to find a place to stay? Not at all. But you will have to be creative. There will be few, if any, hotel rooms available in these eclipse cities by early 2024. Accommodations in the cities and towns along the path of the eclipse will be difficult to come by.

In summer 2017, the author searched on Hotels.com for rooms along the 2017 total eclipse path on the weekend of August 21 and found many major cities sold out. Once word of the 2024 eclipse spreads, room rates will increase and availability will drop.

Search for rooms farther away from the eclipse path. If you are willing to stay in cities outside the eclipse path, you will have better success at finding rooms. As the eclipse approaches, people will book rooms farther from the totality path. By early spring, rooms in cities near the total eclipse path may be unavailable. The effect of this event will be felt across Missouri and the rest of the United States.

Think regionally when looking for rooms. Be prepared to search far and wide during this major event. If a five-hour drive is manageable, your lodging options greatly expand, but it also increases your travel risk.

Internet Rentals

To find rooms to stay in towns along the eclipse path, try a web service such as Airbnb.com. Note that some people rent out rooms or homes illegally, against zoning regulations. Cities will feel the crunch of inquiries early due to others who experienced the 2017 eclipse.

If cities fully enforce zoning laws, authorities may prevent your weekend home rental. Online home rentals during the eclipse will be a target for rental scams. People from out of the area steal photos and descriptions, then post the home for rent. You send your check or wire money to a "rental agent" then show up to find you have been scammed. If the deal sounds strange or too good to be true, run away.

Camping

If you can book a campsite, do it as soon as you can. Do not wait. All areas in the national forests are first-come, first-served. Forest roads may be packed. Expect all areas to be swarming with people. Show up early to stake out your spot. Consider staying farther away and driving early on April 8.

Please respect private land too. Missouri folks don't take kindly to people overrunning their property without permission. In a big state with millions of residents, people are very protective, but they're friendly, too. You never know what you might be able to arrange with a smile and a bit of money.

This all said, there are plenty of camping opportunities throughout Missouri. You don't have to sleep exactly on the eclipse path. If you're ready to rough it, there are national forest camping options.

Government agencies will meet years in advance to talk about how to manage the influx of people. Every possible government agency will be working full time to enforce the various rules and regulations.

National Parks and Monuments

Finding a camping site at any state park, national park, or national monument along the eclipse path in Missouri will be challenging. To watch the eclipse from any location, you do not have to sleep in it. You just need to drive there in the morning.

Law enforcement will be present on the eclipse weekend. Hundreds of thousands of people are expected in the region. Parking may overflow. It will make parking lots and lines on Black Friday at the mall look uncrowded. For an event of this magnitude, find your location as early as possible.

The first sentence of the national parks mission statement is:

"The National Park Service preserves unimpaired the natural and cultural resources and values of the national park system for the enjoyment, education, and inspiration of this and future generations."

Roadside camping (sleeping in your car) is not allowed in national monuments or parks. Park facilities are only designed to handle so many people per day. Water, trash collection, and toilets can only withstand so much. If you notice trash on the ground, take a moment to throw it away. Protect your national park and help out. Rangers are diligent and hardworking but they can only do so much to manage the expected crowds.

National and State Forests

There are national and state forest options in Missouri. They all have camping opportunities. The forest service manages undeveloped and primitive campsites. Be sure to check for any fire restrictions. Check with individual agencies for last-minute information and regulations. The forest service requires proper food storage. Plan to purchase food and water before choosing your campsite. Below is a partial list of national forests along or near the total eclipse path:

Mark Twain NF: www.fs.usda.gov/main/mtnf
Bell Mountain Wilderness
Rockpile Mountain Wilderness
Mingo National Wildlife Refuge and Wilderness

Backcountry service roads abound in Missouri. Maps for forests are available at local visitor centers and bookstores. This book's website has digital copies of some forest maps.

Printed national forest maps are large and detailed. They have illustrated road paths, connections, and other vital travel information not available on digital device maps. Viewing digital maps on your smartphone or mobile pad is difficult. If you plan to camp in the forest, a real paper map is a wise investment.

Camping in federal wilderness areas is also allowed. Those areas afford the ultimate backcountry experience. However, be aware that no vehicle travel is allowed in the specially designated areas. This ban includes: vehicles, bikes, hang gliders, and drones. You can travel only on foot or with pack animals.

Sleep in Your Car

Countless RVs, campers, trucks, cars, and motorcycles will flood Missouri. Sleeping in your car with friends is tolerable. Doing so with unadventurous spouses or children is another matter.

Do not be caught along the path of the total eclipse without some sort of plan, especially in the bigger cities of Missouri. The whole path of totality will fill with people on April 8.

Useful Local Webcams

Local webcams are handy to make last-minute travel decisions. Modern webcams are sensitive enough to show headlights at night. Use them to determine if there are issues before traveling out. Eclipse traffic will add to the morning commuter traffic.

There are smartphone applications which are useful to check webcams in many locations. Consult your device's app store for the latest updates. Whether you use an app or computer, an Internet search will reveal many handy webcams for your eclipse planning.

Weather

It's all about the weather during the eclipse. Nothing else will matter if the sky is cloudy. You can be nearly anywhere along the path in Missouri and catch a view of the event when traffic comes to a standstill. But if there's a cloud cover forecast, seriously reconsider your viewing location.

Travel early wherever you plan to go. Attempting to change locations an hour before the eclipse due to weather will likely cause you to miss the event. Missouri country roads can be narrow and slow. The number of vehicles will cause unexpected backups.

Modern Forecasts

Use a smartphone application to check the up-to-date weather. Wunderground is a good application and has relatively reliable forecasts for the region. The hourly forecast for the same day has been rather accurate for the last two years. The below discussion refers to features found in the Wunderground app. However, any application with detailed weather views will improve your eclipse forecasting skills.

Cloud Cover Forecast

The most useful forecast view is the visible and infrared cloud-coverage map. Avoid downloading this app the night before and trying to learn how to read it. Practice reading them at home. It's imperative to understand how to interpret the maps early.

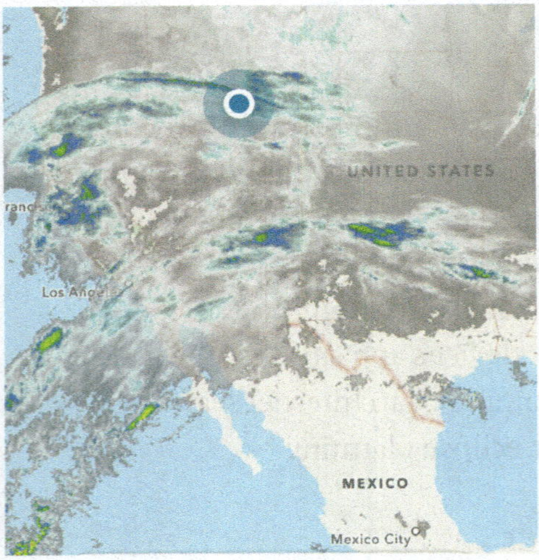

Infrared cloud map showing the worst case eclipse cloud cover. Courtesy of National Weather Service.

All cloud cover, night or day, will appear on an infrared map. Warm, low-altitude clouds are shown in white and gray. High-altitude cold clouds are displayed in shades of green, yellow, red, and purple. Anything other than a clear map spells eclipse-viewing problems.

To improve your weather guess, use the animated viewer of the cloud cover. It will give you a sense of cloud motion. You can discern whether clouds or rain are moving toward, away from, or circulating around your location.

Normal Missouri Weather Pattern

Due to the direction of the jet stream, most weather travels across the Pacific Ocean, through the western states, over the Rockies, and then into Missouri. On occasion, weather can approach from Canada or Mexico. Due to the nature of the storms from the Arctic, weather in Missouri can be unpredictable.

The common weather pattern in April is slightly warm in the afternoon and mildly cool in the evenings. Passing cold fronts in spring can bring unexpected cloud cover and rains.

Historically, Missouri tends to have moderate cloud cover during April. Prepare to make adjustments. If anything other than clear skies are predicted, drive to other parts of Missouri, Arkansas, or Illinois.

Be aware of tornadoes in Missouri. Although the peak tornado season is June, there have been many recorded tornadoes in April. Pay attention to the weather forecast. If dangerous weather is predicted, your main concern should be safety rather than chasing an eclipse.

Consider that slow-moving clouds can obscure the sun for far longer than the four-minute duration of the totality. The time of totality is so short that you do not want to risk it. Missing it due to a single cloud will be a major disappointment.

Local Eclipse Weather Forecasts

Local town and city newspapers, radio, and television stations around Missouri will have a weekend edition with articles discussing the eclipse weather. However, conditions change unpredictably in Missouri. A three-day forecast for April may be incorrect.

Finding the Right Location to View Eclipse Effects

One of the peculiarities of total eclipses is that the entire show is not only in the sky. There are other unusual effects seen during the total eclipse that are worth looking for.

The first effect to watch for is the crescent moon shapes created from leaf shadows on the ground. They're best viewed on a sidewalk or asphalt. They can only be observed during the partial eclipse. The other effect that is worth watching for is the shadow bands or "snakes" as they're commonly called.

Shadow banding is seen right before and after the totality takes place. They're best observed on smooth, plain-colored surfaces. If you plan to be in the forest for the eclipse, you may struggle to see the bands but will likely see crescent shadows all around on the ground.

One of the supreme challenges with all of these effects is choosing what to watch. You can see the crescent shadows in the hour before and after the totality but shadow banding happens before or after totality. It is more difficult to look away from the eclipse than you think.

Road Closures and Traffic

Highways connecting various Missouri towns in the total eclipse path will be heavily impacted on the weekend before and day of the eclipse. As was found in the 2017 total eclipse, there is no way to predict which areas will be impacted.

Planning ahead is essential to give you the best opportunity to enjoy the eclipse without the nightmare of being stuck in traffic for hours on end. The traffic in Oregon and Idaho was stunning, so imagine what it will be like for Missouri.

Update yourself with the latest road report information from the Missouri road condition website:

https://www.modot.org/

It's imperative to plan early and have one if not more backup plans in case of difficult travel conditions. April weather is unpredictable and variable.

If you believe it's necessary to leave a town to watch the eclipse, do so the night before or extremely early in the morning. RVs are common, and trains of them crawl through popular areas.

Communication Information

Cellular Phones

Cellular "cell" phone service in some remote Missouri locations may be problematic. Most of the time there is good coverage along the main highways and interstates. However, even along major thoroughfares, there can be little or no coverage.

It's possible to find zones where text messages will send when phone calls are impossible. If you cannot make a phone call, the chance of having data coverage for web surfing or e-mail is low.

Please look up any information or communicate what you need before departing from the main roads around Missouri. Bureau of Land Management (BLM) areas sometimes have coverage. Planned to be self-contained. Plan for your cell phone not to connect.

You may find yourself out of cell service. With a large number of cell users in a concentrated area, coverage and data speed may collapse as well. Search on the phrase "cell phone coverage breathing".

Wilderness and Forest Safety

All Missouri forest and wilderness areas are full of wild animals. Although beautiful, wild animals can be dangerous. They can easily injure or kill people, as they are far more powerful than humans. Do not try to feed any wild animals, including squirrels, foxes, and chipmunks, as they can carry diseases. These suggestions apply to all public lands.

Ticks

Ticks exist all across the United States, but not all species transmit disease. Ticks cannot fly or jump, but they climb grasses in shrubs in order to attach to people or animals that pass by. Ticks feed on the blood of their host. In doing so, they can transmit potentially life-threatening diseases such as Lyme disease.

Spiders

Although the chance of encountering a venomous spider is low, it is not uncommon to encounter them. The two dangerous spiders in the state are the black widow and brown recluse. Should you encounter either species, simply leave it alone. If you are bitten, seek immediate medical attention, as their toxin can be life-threatening.

Venomous Snakes

There are multiple species of venomous snakes in Missouri including the Copperhead, Cottonmouth, Massasauga Rattlesnake, Western Pygmy Rattlesnake, and Timber Rattlesnake. Although these reptiles are not generally aggressive, they can strike when provoked or threatened. Of the approximately 8,000 people annually bitten by venomous snakes in the United States, ten to fifteen people die according to the U.S. Food and Drug Administration.

The best way to avoid rattlesnake encounters is to be mindful of your environment. Do not place your hands or feet in locations where you cannot clearly see the surroundings. Avoid heavy brush or tall weeds where snakes hide during the day. Step on a log or rock rather than over it, as a hidden snake might be on the other side. Rattlesnakes may not make any noise before striking.

Avoid handling all snakes. Should you be bitten, stay calm and call 911 or emergency dispatch as soon as possible. Transport the victim to the nearest medical facility immediately. Rapid professional treatment is the best way to manage rattlesnake bites. Refer to US Forest Service and professional medical texts for more information on managing rattlesnakes injuries.

Bears

The forests of Missouri are potentially home to black bears. Though they are listed rarely seen, they have been sighted in the state. Although they often appear docile, they can become aggressive if threatened. In the unlikely event of an attack, fight back against the bear. Use whatever you have at your disposal to defend yourself. Report all negative or aggressive bears to the local authorities.

If a bear hears you, it will usually vacate the area. Bear charges are often caused by unexpected and surprise encounters. Noise is the best defense to avoid surprising bears. Regularly clap, make noise, and talk loudly. The Missouri Department of Conservation website has more specific information on safety and food management in bear country at https://nature.mdc.mo.gov/discover-nature/report-wildlife-sightings/bear-reports.

It is recommended to stay one hundred yards (300 feet) away from all bears. They are exciting to see but need their space. Refer to current forest or park regulations for more safety information.

Mountain Lions

Though rare in the state, there have been recent reports of mountain lions by the Missouri Department of Conservation. If you encounter a mountain lion, do not run. Keep calm, back away slowly, and maintain eye contact. Do all you can to appear larger. Stand upright, raise your arms, or hoist your jacket. Never bend over or crouch down. If attacked, fight back.

Eclipse Day Safety

1. Hydrate

Spring temperatures are usually mild to warm. The excitement of the event can distract you from managing hydration. Drink plenty of water. Consume more than you would at home.

2. Eye Safety time

Use certified eclipse safety glasses at all times when viewing the partial eclipse. Only remove the glasses when the totality happens. Give your eyes time to rest. They can dry out and become irritated. Bring FDA approved eye drops to keep your eyes moist.

3. Sun exposure

Facing at the sun for three hours can result in sunburns. Wear sunglasses and liberally apply sunscreen to avoid sunburns.

4. Eat well

Keep your energy up. Appetite loss is common when traveling. Maintain your normal eating schedule.

5. Prepare for temperature changes

Temperatures will drop rapidly during the eclipse and also once the sun sets. Bring appropriate clothing.

6. Talk with your doctor

If the humidity or heat bothers you talk with your doctor before traveling. Seek professional medical attention for serious symptoms.

Total eclipse path across the United States (approximate).

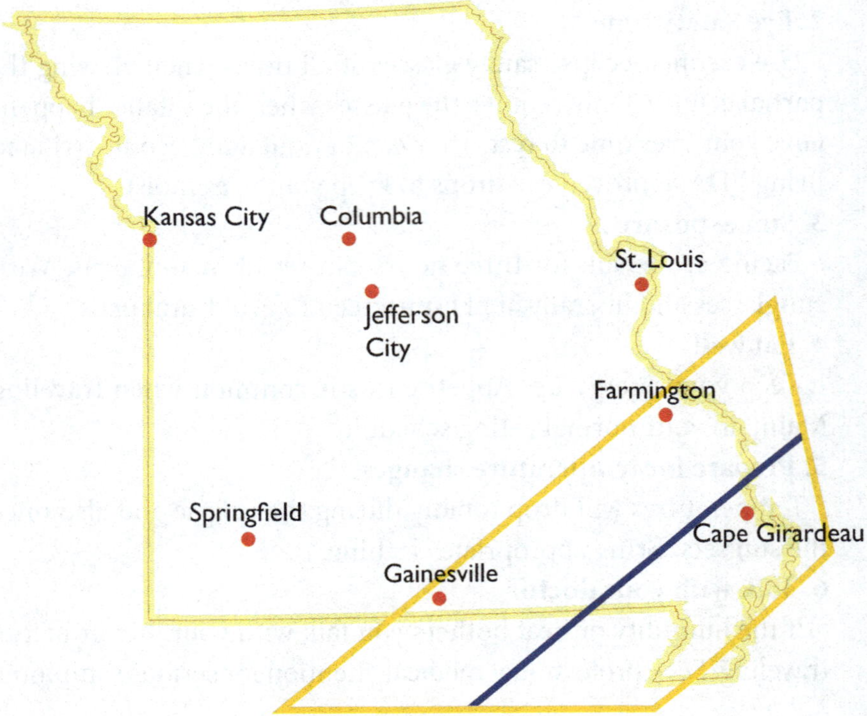

Total eclipse path across Missouri (approximate).

All About Eclipses

How an Eclipse Happens

An eclipse occurs when one celestial body falls in line with another, thus obscuring the sun from view. This occurs much more often than you'd think, considering how many bodies there are in the solar system. For instance, there are over 150 moons in the solar system. On Earth, we have two primary celestial bodies: the sun and the moon. The entire solar system is constantly in motion, with planets orbiting the sun and moons orbiting the planets. These celestial bodies often come into alignment. When these alignments cause the sun to be blocked, it is called an eclipse.

For an eclipse to occur, the sun, Earth, and moon must be in alignment. There are two types of eclipses: solar and lunar. A solar eclipse occurs when the moon obscures the sun. A lunar eclipse occurs when the moon passes through Earth's shadow. Solar eclipses are much more common, as we experience an average of 240 solar eclipses a century compared to an average of 150 lunar eclipses. Despite this, we are more likely to see a lunar eclipse than a solar eclipse. This is due to the visibility of each.

For a solar eclipse to be visible, you have to be in the moon's shadow. The problem with viewing a total eclipse is that the moon casts a small shadow over the world at any given time. You have to be in

* ILLUSTRATION NOT TO SCALE

a precise location to view a total eclipse. The issue that arises is that most of these locations are inaccessible to most people. Though many would like to see a total solar eclipse, most aren't about to set sail for the middle of the Pacific Ocean. In fact, a solar eclipse is visible in the same place on the world on average every 375 years. This means that if you miss a solar eclipse above your hometown, you're not going to see another one unless you travel or move.

It's much easier to catch a glimpse of a lunar eclipse, even though they occur at a much lower frequency than their solar counterparts. A lunar eclipse darkens the moon for a few hours. This is different than a new moon when it faces away from the sun. During these eclipses, the moon fades and becomes nearly invisible.

Another result of a lunar eclipse is a blood moon. Earth's atmosphere bends a small amount of sunlight onto the moon turning it orange-red. The blood moon is caused by the dawn or dusk light being refracted onto the moon during an eclipse.

Lunar eclipses are much easier to see. Even when the moon is in the shadow of Earth, it's still visible throughout the world because of how much smaller it is than Earth.

Total vs. Partial Eclipse

What is the difference between a partial and total eclipse? A total eclipse of either the sun or the moon will occur only when the sun, Earth, and the moon are aligned in a perfectly straight line. This ensures that either the sun or the moon is partially or completely obscured.

In contrast, a partial eclipse occurs when the alignment of the three celestial bodies is not in a perfectly straight line. These types of eclipses usually result in only a part of either the sun or the moon being obscured. This is often what led to ancient civilizations believing that some form of magical beast or deity was eating the sun or the moon. It appears as though something has taken a bite out of either the sun or the moon during a partial eclipse.

Total eclipses, rarer than partial eclipses, still occur quite often. It's more difficult for people to be in a position to experience such an event firsthand. Total solar eclipses can only be viewed from a small portion of the world that falls into the darkest part of the moon's shadow. Often this happens in the middle of the ocean.

The Moon's Shadow

The moon's shadow is divided into two parts: the umbra and the penumbra. The former is much smaller than the latter, as the umbra is the innermost and darkest part of the shadow. The umbra is thus the central point of the moon's shadow, meaning that it is extremely small in comparison to the entire shadow. For a total solar eclipse to be visible, you need to be directly beneath the umbra of the moon's shadow. This is because that is the only point at which the moon completely blocks the view of the sun.

In contrast, the penumbra is the region of the moon's shadow in which only a portion of the light cast by the sun is obscured. When

Total eclipse shadow 2016 as seen from 1 million miles on the Deep Space Climate Observatory satellite. Courtesy of NASA.

standing in the penumbra, you are viewing the eclipse at an angle. In the penumbra, the moon does not completely block the sun from view. This means that while the event is a total solar eclipse, you'll only see a partial eclipse. The umbra for the April 8 eclipse is over one hundred miles wide. The penumbra will cover much of the United States.

To provide some context, one total solar eclipse we experienced occurred on March 9, 2016, and was visible as a partial eclipse across most of the Pacific Ocean, parts of Asia, and Australia. However, the only place in the world to view this total solar eclipse was in a few parts of Indonesia.

Due to the varied locations and brief periods for which they're visible, it's difficult to see each and every eclipse that occurs. The umbra of the moon is such a small fraction of the entire shadow and the majority of our planet is comprised of water. The rarity of being able to view a total solar eclipse increases significantly because it's likely that the umbra will fall over some part of the ocean rather than a populated landmass. There are not total or annular eclipses every month because the moon's orbit is 5.1° off the ecliptic plane of the Earth and sun.

Eclipses Throughout History

Ancient peoples believed eclipses were from the wrath of angry gods, portents of doom and misfortune, or wars between celestial beings. Eclipses have played many roles in cultures, creating myths since the dawn of time. Both solar and lunar eclipses affected societies worldwide. Inspiring fear, curiosity, and the creation of legends, eclipses have cast a long shadow in the collective unconscious of humanity throughout history.

Early Myth & Astronomy

Documented observations of solar eclipses have been found as far back in history as ancient Egyptian and Chinese records. Timekeeping was important to ancient Chinese cultures. Astronomical observations were an integral factor in the Chinese calendar. The first observation of a solar eclipse is found in Chinese records from over 4,000 years ago. Evidence suggests that ancient Egyptian observations may predate those archaic writings.

Many ancient societies, including Roman, Greek and Chinese civilizations, were able to infer and foresee solar eclipses from astronomical data. The sudden and unpredictable nature of solar eclipses had a stressful and intimidating effect on many societies that lacked the scientific insight to accurately predict astronomical events. Relying on the sun for their agricultural livelihood, those societies interpreted solar eclipses as world-threatening disasters.

In ancient Vietnam, solar eclipses were explained as a giant frog eating the sun. The peasantry of ancient Greece believed that an eclipse was the sign of a furious godhead, presenting an omen of wrathful retribution in the form of natural disasters. Other cultures were less speculative in their investigations. The Chinese Song Dynasty scientist Shen Kuo proved the spherical nature of the Earth and heavenly bodies through scientific insight gained by the study of eclipses.

The Eclipse in Native American Mythology

Eclipses have played a significant role in the history of the United States. Before Europeans settled in the Americas, solar eclipses were important astronomical events to Native American cultures. In most native cultures, an eclipse was a particularly bad omen. Both the sun and the moon were regarded as sacred. Viewing an eclipse, or even being outside for the duration of the event, was considered highly taboo by the Navajo culture. During an eclipse, men and women would simply avert their eyes from the sky, acting as though it was not happening.

The Choctaw people had a unique story to explain solar eclipses. Considering the event as the mischievous actions of a black squirrel and its attempt to eat the sun, the Choctaw people would do their best to scare away the cosmic squirrel by making as much noise as possible until the end of the event, at which point cognitive bias would cause them to believe they'd once again averted disaster on an interplanetary scale.

Contemporary American Solar Phenomena

The investigation of solar phenomena in twentieth-century American history had a similarly profound effect on the people of the United States. A total solar eclipse occurring on the sixteenth of June, 1806, engulfed the entire country. It started near modern-day Arizona. It passed across the Midwest, over Ohio, Pennsylvania, New York, Massachusetts, and Connecticut. The 1806 total eclipse was notable for being one of the first publicly advertised solar events. The public was informed beforehand of the astronomical curiosity through a pamphlet written by Andrew Newell entitled *Darkness at Noon, or the Great Solar Eclipse*.

This pamphlet described local circumstances and went into great detail explaining the true nature of the phenomenon, dispelling myth and superstition, and even giving questionable advice on the best methods of viewing the sun during the event. Replete with a short historical record of eclipses through the ages, the *Darkness at Noon* pamphlet is one of the first examples of an attempt to capitalize on the mysterious nature of solar eclipses.

Another notable American solar eclipse occurred on June 8, 1918. Passing over the United States from Washington to Florida, the eclipse was accurately predicted by the U.S. Naval Observatory and heavily documented in the newspapers of the day. Howard Russell Butler, painter and founder of the American Fine Arts Society, painted the eclipse from the U.S. Naval Observatory, immortalizing the event in *The Oregon Eclipse*.

Four more total solar eclipses occurred over the United States in the years 1923, 1925, 1932, and 1954, with another occurring in 1959. The October 2, 1959, solar eclipse began over Boston, Massachusetts. It was a sunrise event that was unviewable from the ground level. Eminent astronomer Jay Pasachoff attributed this event to sparking his interest in the study of astronomy. Studying under Professor Donald Menzel of Williams College, Pasachoff was able to view the event from an airline hired by his professor.

To this day, many myths surround the eclipse. In India, some local customs require fasting. In eastern Africa, eclipses are seen as a danger to pregnant women and young children. Despite the mystery and legend associated with unique and rare astronomical events, eclipses continue to be awe-inspiring. Even in the modern day, eclipses draw out reverential respect for the inexorable passing of celestial bodies. They are a reminder of the intimate relationship between the denizens of Earth and the universe at large.

Present Day Eclipses

The year 2017 brought the world's most-watched total eclipse in history on August 21, when a total solar eclipse crossed the United States. An annular eclipse, a "ring of fire," will pass over the United States in 2023 from Oregon to Texas. Though impressive, it will not compare to the 2024 total eclipse. There is little in nature that equals

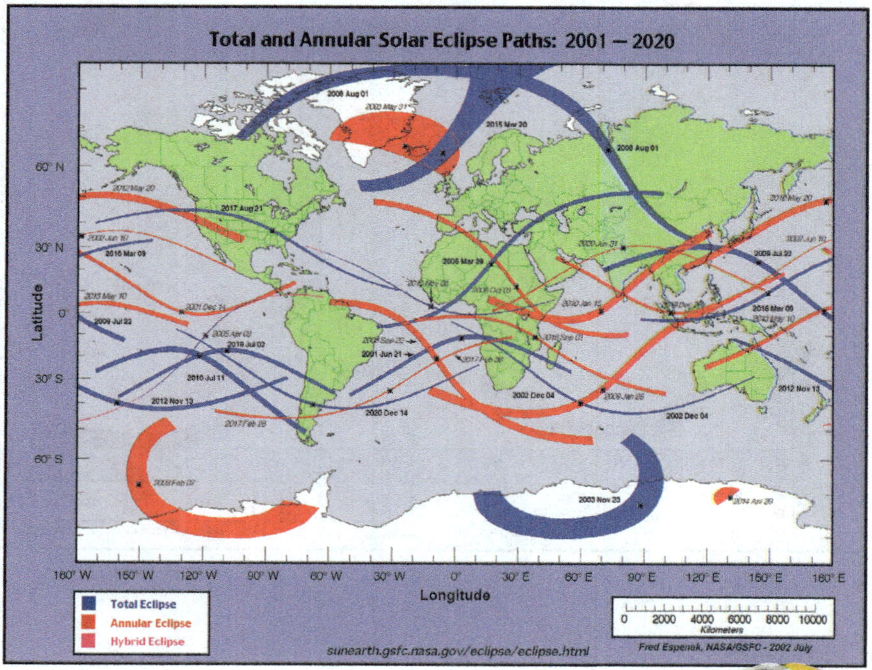

COURTESY OF NASA

the spectacle of the sun's corona and seeing stars in the day.

There will be multiple partial, annular, or hybrid eclipses across the world before the 2024 total eclipse. However many are in remote, inaccessible, or potentially dangerous locations on the globe. In 2019 and 2020, Chile and Argentina will experience total eclipses. The next total eclipse after that will occur over Antarctica in 2021. An extremely rare hybrid eclipse will happen in 2023 over the Indian Ocean, Australia, and Indonesia.

The next total solar eclipse viewable from the United States will occur on April 8, 2024. It will be visible in fifteen states: Texas, Oklahoma, Arkansas, Missouri, Tennessee, Kentucky, Illinois, Indiana, Ohio, Pennsylvania, Michigan, New York, Vermont, New Hampshire, and Maine.

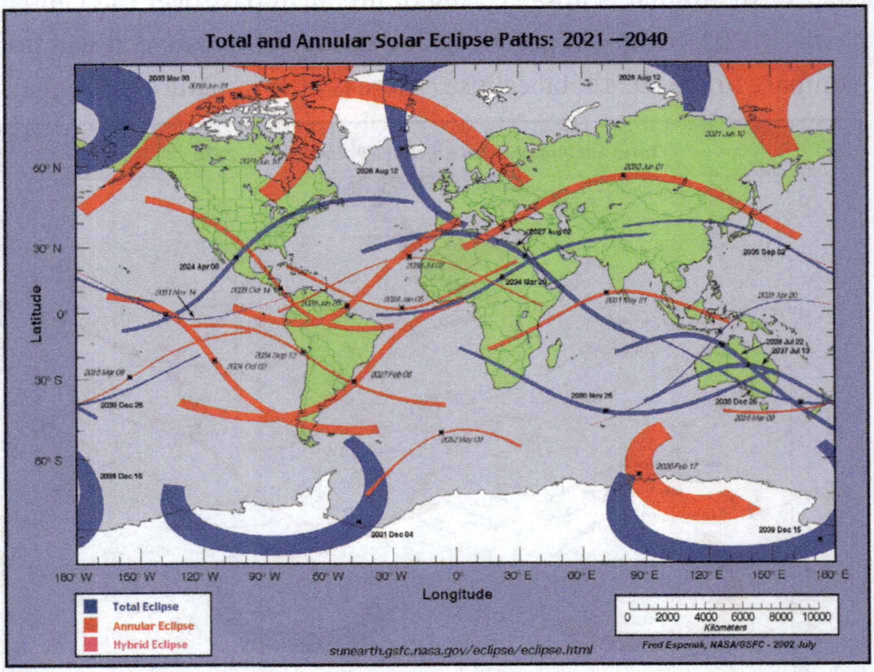

COURTESY OF NASA

Viewing and Photographing the Eclipse

At-home Pinhole Method

Use the pinhole method to view the eclipse safely. It costs little but is the safest technique there is. Take a stiff piece of single-layer cardboard and punch a clean pinhole. Let the sun shine through the pinhole onto another piece of cardboard. That's it!

Never look at the sun through the pinhole. Your back should be toward the sun to protect your eyes. To brighten the image, simply move the back piece of cardboard closer to the pinhole. To see it larger, move the back cardboard farther away. Do not make the pinhole larger. It will only distort the crescent sun.

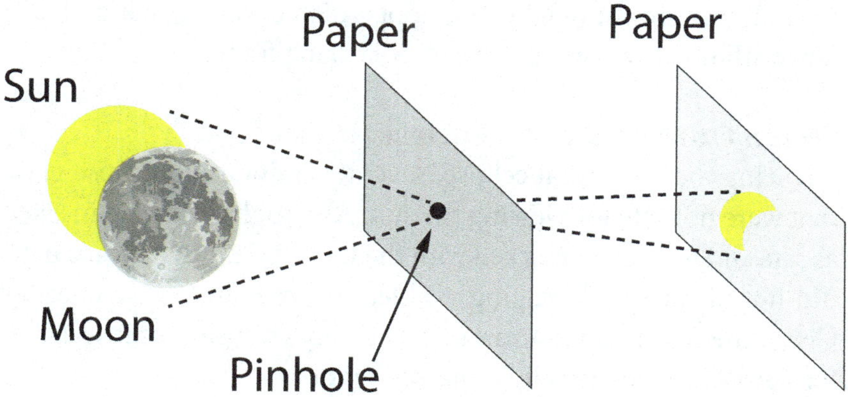

Welding Goggles

Welding goggles that have a rating of fourteen or higher are another useful eclipse viewing tool. The goggles can be used to view the solar eclipse directly. Do not use the goggles to look through binoculars or telescopes, as the goggles could potentially shatter due to intense direct heat. Avoid long periods of gazing with the goggles. Look away every so often. Give your eyes a break.

Solar Filters for Telescopes

The ONLY safe way to view solar eclipses using telescopes or binoculars is to use solar filters. The filters are coated with metal

to diminish the full intensity of the sun. Although the filters can be expensive, it is better to purchase a quality filter rather than an inexpensive one that could shatter or melt from the heat.

The filters attach to the front of the telescope for easy viewing. Remember to give your telescope cooling breaks. Rapid heating can damage your equipment with or without filters attached.

Watch Out for Unsafe Filters

There are several myths surrounding solar filters for eclipse viewing. In order for filters to be safe, they must be specially designed for looking at a solar eclipse. The following are all unsafe for eclipse viewing and can lead to retinal damage: developed colored or chromogenic film, black-and-white negatives such as X-rays, CDs with aluminum, smoked glass, floppy disk covers, black-and-white film with no silver, sunglasses, or polarizing films.

Watch Out for Unsafe Eclipse Glasses

During the 2017 total eclipse, several vendors sold eclipse glasses that were not safe for viewing the sun. Although they were marketed as safe and were even marked with the ISO 12312-2 certification, they did not block eye-damaging visible, infrared, and ultraviolet light. Check the American Astronomical Society's website (eclipse.aas.org) for a list of reputable eclipse glasses vendors.

Viewing with Binoculars

When viewing the eclipse with binoculars, it is important to use solar filters on both lenses until totality. Only then is it safe to remove the filter. As the sun becomes visible after totality, replace the filters for safe viewing. Protect your pupils. Remember to give your binoculars a cool-down break between viewings. They can overheat rapidly from being pointed directly at the sun even with filters attached.

Planning Ahead

There are many things to keep in mind when viewing a total eclipse. It is important to plan ahead to get the most out of this extraordinary experience.

Understanding Sun Position

All compass bearings in this book are true north. All compasses point to Earth's magnetic north. The difference between these two measurements is called magnetic declination. The magnetic declination for Missouri is:

0° 51' E ± 0° 37' (for Jefferson City in 2024)

Adjust the declination from the azimuth bearing as given in the text, and set your compass to that direction.

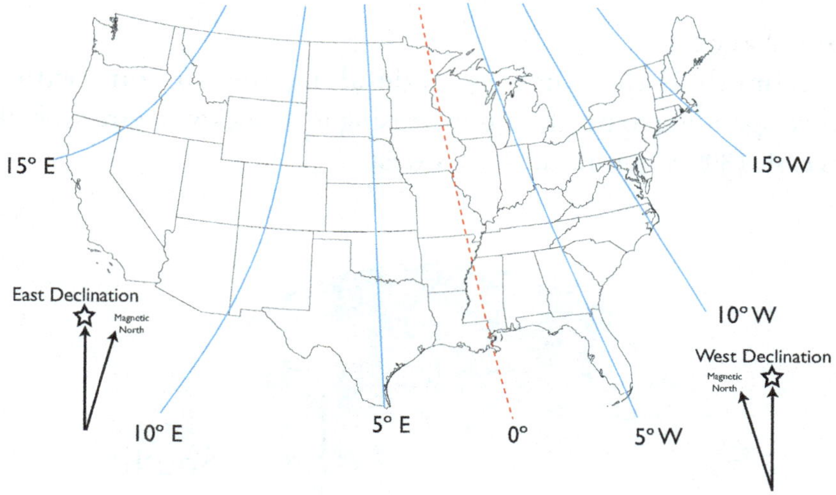

If you purchase a compass with a built-in declination adjustment, you can change the setting once and eliminate the calculations. The Suunto M-3G compass has this correction. A compass with a sighting mirror or wire will help you make a more accurate azimuth sighting.

The Suunto M-3G also has an inclinometer. This allows you to measure the elevation of any object above the horizon. Use this to figure out how high the sun will be above your position.

You can also use a smartphone inclinometer and compass for this purpose. Make sure to calibrate your smartphone's compass before every use, otherwise it might indicate the wrong bearing. Set the smartphone compass for true north to match the book. Understand the compass prior to April 8. There will be little time to guess or search on the Internet. Smartphone and GPS compasses are "sticky." Their compasses don't swing as freely as a magnetic compass does.

The author has used his magnetic compass for azimuth measurements and a smartphone to measure elevation. Combining these two tools will allow you to make the best sightings possible.

Outdoor sporting goods stores in most towns and cities carry compasses. Purchase and practice with a good compass in your hometown well before the event. Take the time to learn how to use it before the day of the eclipse. You do not want to struggle with orienteering basics under pressure.

Sun Azimuth

Azimuth is the compass angle along the horizon, with 0° corresponding to north, and increasing in a clockwise direction. 90° is east, 180° is south, and 270° is west.

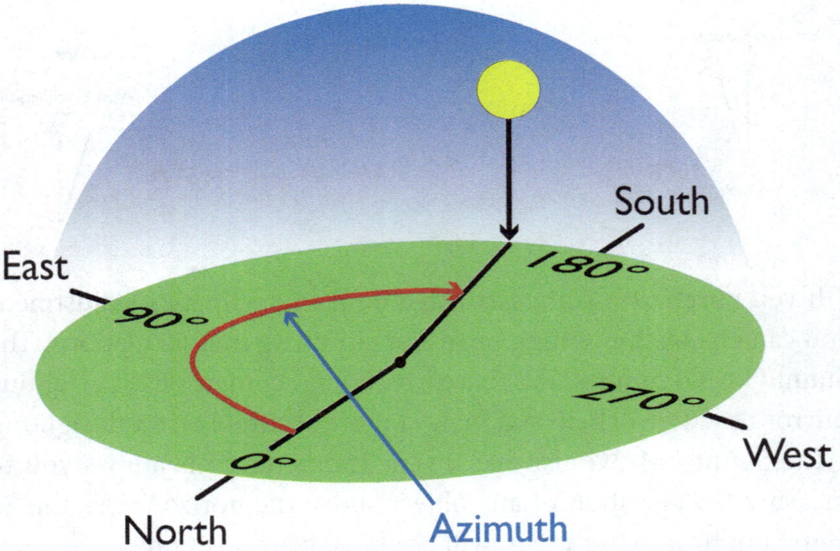

Sun Elevation

Altitude is the sun's angle up from the horizon. A 0° altitude means exactly on the horizon and 90° means "straight up."

Using the sun azimuth and elevation data, you can predict the position of the sun at any given time. Positions given in this book coincide with the time of eclipse totality unless otherwise noted.

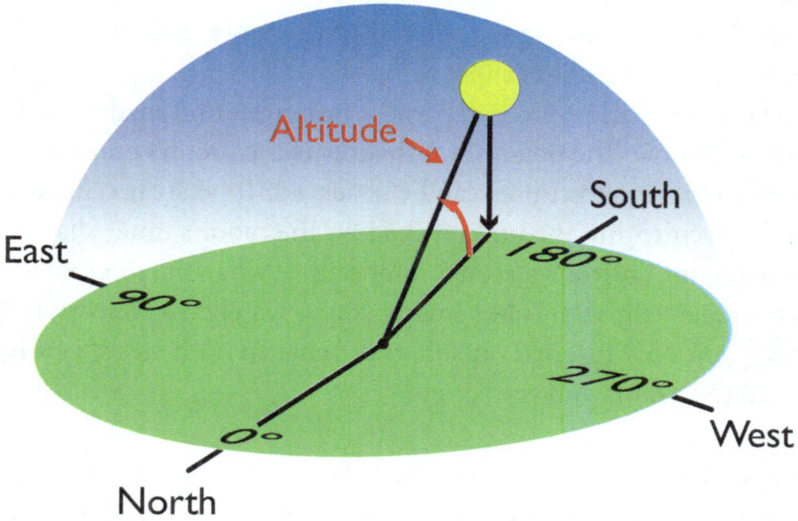

Eclipse Data for Select Missouri Locations

LOCATION	TOTALITY START (CDT)	ALTITUDE	AZIMUTH
CAPE GIRARDEAU	1:58:19PM	57°	207°
FARMINGTON	1:58:15PM	57°	205°
KENNETT	1:57:10PM	58°	207°
POPLAR BLUFF	1:56:18PM	58°	205°
SIKESTON	1:58:00PM	57°	207°
WEST PLAINS	1:54:31PM	59°	202°

Eclipse Photography

Photographing an eclipse is an exciting challenge, as the moon's shadow moves near 1,600MPH. There is an element of danger and the pressure of time. Looking at the unfiltered sun through a camera can permanently damage your vision and your camera. If you are unsure, just enjoy the eclipse with specially designed eclipse glasses. Keep a solar filter on your lens during the eclipse and remove for the duration of totality!

Partial Vs. Total Solar Eclipse

To successfully and safely photograph a partial and total eclipse, it is important to understand the difference between the two. A solar eclipse occurs when the moon is positioned between the sun and Earth. The region where the shadow of the moon falls upon Earth's surface is where a solar eclipse is visible.

The moon's shadow has two parts—the penumbral shadow and the umbral shadow. The penumbral shadow is the moon's outer shadow where partial solar eclipses can be observed. Total solar eclipses can only be seen within the umbral shadow, the moon's inner shadow.

You cannot say you've seen a total eclipse when all you saw was a partial solar eclipse. It is like saying you've watched a concert, but in reality, you only listened outside the arena. In both cases, you have missed the drama and the action.

Photographing A Partial And Total Solar Eclipse

Aside from the region where the outer shadow of the moon is cast, a partial solar eclipse is also visible before a total solar eclipse within

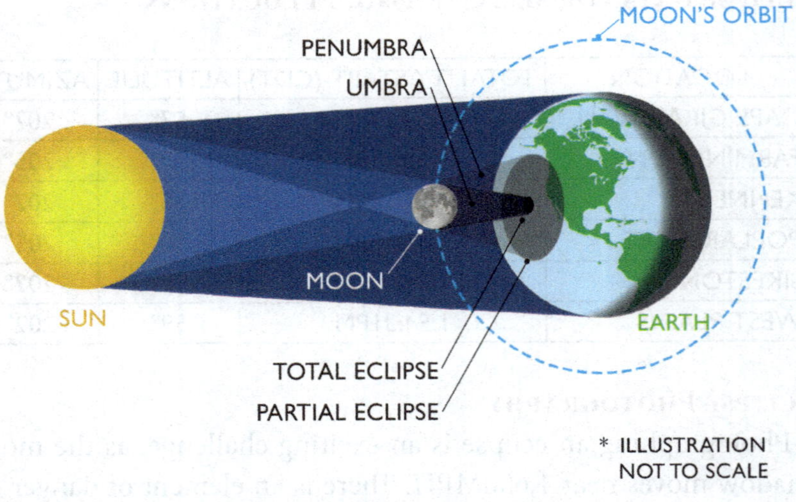

the inner shadow region. In both cases, it is imperative to use a solar filter on the lens for both photography and safety reasons. This is the only difference between taking a partial eclipse and a total eclipse photograph of the sun.

To photograph a total solar eclipse, you must be within the Path of Totality, the surface of the Earth within the moon's umbral shadow.

The Challenge

A total solar eclipse only lasts for a couple of minutes. It is brief, but the scenario it brings is unforgettable. Seeing the radiant sun slowly being covered by darkness gives the spectator a high level of anticipation and indescribable excitement. Once the moon completely covers the sun's radiance, the corona is finally visible. In the darkness, the sun's corona shines, capturing the crowd's full attention. Watching this phenomenon is a breathtaking experience.

Amidst all the noise, cheering, and excitement, you have no more than a few minutes to take a perfect photograph. The key to this is planning. You need to plan, practice, and perfect what you will do when the big moment arrives because there is no replay. The pressure is enormous. You only have a short time to capture the totality and the sun's corona using different exposures.

Plan, Practice, Perfect

It is important to practice photographing before the actual phenomenon arrives. Test your chosen imaging setup for flaws. Rehearse over and over until your body remembers what you will do from the moment you arrive at your chosen spot to the moment you pack up and leave the area.

You will discover potential problems regarding vibrations and focus that you can address immediately. This minimizes the variables that might affect your photographs at the most critical moment.

It's common for experienced eclipse chasers to lose track of what they plan to do. Write down what you expect to do. Practice it time and again. Play annoying, distracting music while you practice. Try photographing in the worst weather possible. Do anything you can to practice under pressure. Eclipse day is not the time to practice.

Once the sun is completely covered, don't just take photographs. Capture the experience and the image of the total solar eclipse in your mind as well. Set up cameras around you to record not just the total solar eclipse but also the excitement and reaction of the crowd.

Eclipse Photography Gear

What do you need to photograph the total eclipse? There are only a few pieces of equipment that you'll need. Preparing to photograph an eclipse successfully takes time. Not only do you have to be skilled and have the right gear, you have to be in the correct place.

Basic Eclipse Photography Equipment

- Solar viewing glasses (verify authenticity)
- Lens solar filter
- Minimum 300mm lens
- Stable tripod that can be tilted to 60° vertical
- High-resolution DSLR
- Spare batteries for everything
- Secondary camera to photograph people, the horizon, etc.
- Remote cable or wireless release

Additional Items

- Video camera
- Video camera tripod
- Quality pair of binoculars
- Solar filters for each binocular lens
- Photo editing software

Equipment to Prepare Before the Big Day

A. Solar viewing glasses

You need a pair of solar viewing glasses as the eclipse approaches.

B. Solar Filter

Partial and total eclipse photography is different from normal photography. Even if only 1% of the sun's surface is visible, it is still approximately 10,000 times brighter than the moon. Before totality, use a solar filter on your lens. Do not look at the sun with your eyes. It can cause irreparable damage to your retinas.

DO NOT leave your camera pointed at the sun without a solar filter attached. The sun will melt the inside of your camera. Think of a magnifying glass used to torch ants and multiply that by one hundred.

C. Lens

To capture the corona's majesty, you need to use a telescope or a telephoto lens. The best focal length, which will give you a large image of the sun's disk, is 400mm and above. You don't want to waste all your efforts by bringing home a small dot where the black disk and majestic corona are supposed to be.

D. Tripod

Bring a stable enough tripod to support your camera properly to avoid unsteady shots and repeated adjustments. Either will ruin your photos. It also needs to be portable in case you need to change locations for a better shot. ***Shut off camera stabilization on a tripod!***

E. Camera

You need to remember to set your camera to its highest resolution to capture all the details. Set your camera to:

- 14-bit RAW is ideal, otherwise
- JPG, Fine compression, Maximum resolution

Bracket your exposures. Shoot at various shutter speeds to capture different brightnesses in the corona. Note that stopping your lens all the way down may not result in the sharpest images.

Choose the lowest possible ISO for the best quality while maintaining a high shutter speed to prevent blurred shots. Set your camera to manual. Do not use AUTO ISO. Your camera will be fooled. The night before, test the focus position of your lens using a bright star or the moon.

Constantly double-check your focus. Be paranoid about this. You can deal with a grainy picture. No amount of Photoshop will fix a blurry, out-of-focus picture.

F. Batteries

Remember to bring fresh batteries! Make sure that you have enough power to capture the most important moments. Swap in fresh batteries thirty minutes before totality.

G. Remote release

Use a wired or wireless remote release to fire the camera's shutter. This will reduce the amount of camera vibration.

H. Video Camera

Run a video camera of yourself. Capture all the things you say and do during the totality. You'll be amazed at your reaction.

I. Photo editing software

You will need quality photo editing software to process your eclipse images. Adobe Lightroom and Photoshop are excellent programs to extract the most out of your images. Become well versed in how to use them at least a month before the eclipse.

J. Smartphone applications

The following smartphone applications will aid in your photography planning: Wunderground, Skyview, Photographer's Ephemeris, Sunrise and Sunset Calculator, SunCalc, and Sun Surveyor among others.

Camera Phones

Smartphone cameras are useful for many things but not eclipse photography. An iPhone 6 camera has a 63° horizontal field of view and is 3264 pixels across. If you attempt to photograph the eclipse, the sun will be a measly 30-40 pixels wide depending on the phone. Digital pinch zoom won't help here. If you want *National Geographic* images, you'll need a serious camera and lens, far beyond any smartphone.

Consider instead using a smartphone to run a time-lapse of the entire event. The sun will be minuscule when shot on a smartphone. Think of something else exciting and interesting do to with it. Purchase a Gorilla Pod, inexpensive tripod, or selfie stick and mount the smartphone somewhere unique.

Also, partial and total eclipse light is strange and ethereal. Consider using that light to take unique pictures of things and people. It's rare and you may have something no one else does.

Focal Length & the Size of Sun

The size of the sun in a photo depends on the lens focal length. A 300mm lens is the recommended minimum on a full-frame (FF) DSLR. Lenses up to this size are relatively inexpensive. For more magnification, use an APS-C (crop) size sensor. Cameras with these sensors provide an advantage by capturing a larger sun.

For the same focal length, an APS-C sensor will provide a greater apparent magnification of any object. As a consequence, a shorter, less expensive lens can be used to capture the same size sun.

The below figure shows the size of the sun on a camera sensor at various focal lengths. As can be seen with the 200mm lens, the sun is quite small. On a full-frame camera at 200mm, the sun will be 371 pixels wide on a Nikon D810, a 36-megapixel body. A lower resolution FF camera will result in an even smaller sun.

Printing a 24-inch image shot on a Nikon D810 with a 200mm lens at a standard 300 pixels per inch results in a small sun. On this size paper, the sun will be a miserly 1.25 inches wide!

Photographing the eclipse with a lens shorter than 300mm will leave you with little to work with. Using a 400mm lens and printing a 24-inch print will result in a 2.5-inch-wide sun. For as massive as the sun is, it is a challenge to take a large photograph of the sun. The sun will appear to move fast with a 500mm lens, too. Plan to adjust.

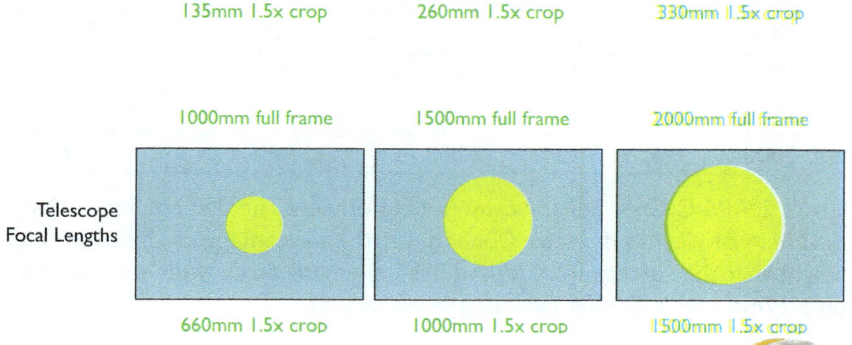

FOCAL LENGTH	FOV FULL FRAME	FF VERT. ANGLE	% OF FF	SUN PIXEL SIZE
14	104° X 81°	81°	0.7%	32.1
20	84° X 62°	62°	0.9%	41.9
28	65° X 46°	46°	1.2%	56.5
35	54° X 38°	38°	1.4%	68.5
50	40° X 27°	27°	2.0%	96.4
105	19° X 13°	13°	4.1%	200.2
200	10° X 7°	7°	7.6%	371.9
400	5° X 3.4°	3.4°	15.6%	765.6
500	4° X 2.7°	2.7°	19.6%	964.2
1000	2° X 1.3°	1.3°	40.8%	2002.5
1500	1.4° X 0.9°	0.9°	58.9%	2892.6
2000	1° X 0.68°	0.68°	77.9%	3828.4

Chart 1: Full-frame camera field of view. The 3rd column is the vertical field of view in degrees. Column 4 is the percentage of the total sensor height that the sun covers. Column 5 is how many pixels wide the sun will be on a 36MP Nikon D810. (Values are estimates)

PHOTOGRAPHY

FOCAL LENGTH	FOV CROP	CROP VERT DEG	% OF CROP	SUN PIXEL SIZE
14	80° X 58°	58°	0.9%	33.9
20	61° X 43°	43°	1.2%	45.8
28	45° X 31°	31°	1.7%	63.5
35	37° X 25°	25°	2.1%	78.7
50	26° X 18°	18°	2.9%	109.3
105	13° X 8°	8°	6.6%	245.9
200	6.7° X 4.5°	4.5°	11.8%	437.2
400	3.4° X 2°	2°	26.5%	983.7
500	2.7° X 1.8	1.8°	29.4%	1093.0
1000	1.3° X 0.9°	0.9°	58.9%	2186.0
1500	0.9° X 0.6°	0.6°	88.3%	3278.9
2000	0.6° X 0.45°	0.5°	117.8%	4371.9

Chart 2: APS-C Crop sensor camera field of view. The 3rd column is the vertical field of view in degrees. Column 4 is the percentage of the total sensor height that the sun covers. Column 5 is how many pixels wide the sun will be on a 12mp Nikon D300s. (Values are estimates)

The big challenge is the cost of the lens. Lenses longer than 300mm are expensive. They also require heavier tripods and specialized tripod heads. The 70-300mm lenses from Nikon, Canon, Tamron, and others are relatively affordable options. It is worth spending time at a local camera shop to try different lenses. Long focal-length lenses are a significant investment, especially for a single event.

To achieve a large eclipse image, you will need a long focal-length lens, ideally at least 400mm. A standard 70-300mm lens set to 300mm will show a small sun. At 500mm, the sun image becomes larger and covers more of the sensor area. The corona will take up a significant portion of the frame. By 1000mm, the corona will exceed the capture area on a full-frame sensor. See the picture in this chapter for sun size simulations for different focal lengths.

Suggested Exposures

To photograph the partial eclipse, the camera must have a solar filter attached. If not, the intense light from the sun may damage (fry) the inside of your camera. This has happened to the author. The exposure depends on the density (darkness) of the solar filter used.

As a starting point, set the camera to ISO 100, f/8, and with the solar filter on, try an exposure of 1/2000. Make adjustments based on the filter used, histogram, and highlight warning.

Turn on the highlight warning in your camera. This feature is commonly called "blinkies." This warning will help you detect if the image is overexposed or not.

Once the Baily's Beads, prominences, and corona become visible, there will only be a few minutes to take bracketed shots. It will take at least eleven shots to capture the various areas of the sun's corona and stars. The brightness varies considerably. No commercially available camera can capture the incredible dynamic range of the different portions of the delicate corona. This requires taking multiple photographs and digitally combining them afterward.

During totality, try these exposure times at ISO 100 and f/8:
1/4000, 1/2000, 1/1000, 1/250, 1/60, 1/30, 1/15, 1/4, 1/2, 1 sec, and 4 sec.

Disable camera/lens stabilization on a tripod!

Photography Time

Set the camera to full-stop adjustments. It will reduce the time spent fiddling. As an example, the author tried the above shot sequence, adjusting the shutter speed as fast as possible.

It took thirty-three seconds to shoot the above 11 shots using 1/3-stop increments. This was without adjusting composition, focus, or anything else but the shutter speed. When the camera was set to full stop increments, it only took twenty-two seconds to step through the same shutter speed sequence. Use a remote release to reduce camera shake.

Assuming the totality lasts less than two minutes, only four shot sequences could be made using 1/3-stop increments. Yet six shot sequences could be made when the camera was set to full stop steps. Zero time was spent looking at the back LCD to analyze highlights and the histogram.

Now add in the bare minimum time to check the highlight warning. It took sixty-three seconds to shoot and check each image using full stops. And that was without changing the composition to allow for sun movement, bumping the tripod, etc. Looking at the LCD ("chimping") consumed **half** of the totality time.

This test was done in the comfort of home under no pressure. In real world conditions, it may be possible to successfully shoot only one sequence. If you plan to capture the entire dynamic range of the totality, you must practice the sequence until you have it down cold. If you normally fumble with your camera, do not underestimate the difficulty, frustration, and stress of total eclipse photography.

Most importantly, trying to shoot this sequence allowed for zero time to simply look at the totality to enjoy the spectacle.

Avoid Last Minute Purchases

You should purchase whatever you think you'll need to photograph the eclipse early. This event will be nothing short of massive. Remember the hot toy of the year? Multiply that frenzy by a thousand. Everyone will want to try to capture their own photo.

Do not wait until the last few weeks before the eclipse to purchase cameras, lenses, filters, tripods, viewing glasses, and associated material. Consider that the totality of the eclipse will streak across

America. Everyone who wants to photograph the eclipse will order at the same time. If you wait until too late to buy what you need, it's conceivable that solar filters to create a total eclipse photo will be sold out in the United States. All filters sold out during the 2017 total eclipse. Whether this happens or not, do not wait to make your purchases. It may be too late.

Practice

You will need to practice with your equipment. Things may go wrong that you don't anticipate. If you've never photographed a partial or total eclipse, taking quality shots is more difficult than you think. Practice shooting the sequence with a midday sun. This will tell you if you have your exposures and timing correct. Figure out what you need well in advance.

Practice photographing the full moon and stars at night. Capture the moon in full daylight to learn how your camera reacts. Astrophotography is challenging and requires practice.

The August 21, 2017, eclipse as seen in Jackson, WY, shot with a Nikon D800 with an 80-400mm lens set to 340mm. The sun is 644 pixels wide on the 7360x4912 image.

This image is shown straight out of the camera without modification. Even with a high-quality camera and lens, photographing an eclipse is challenging.

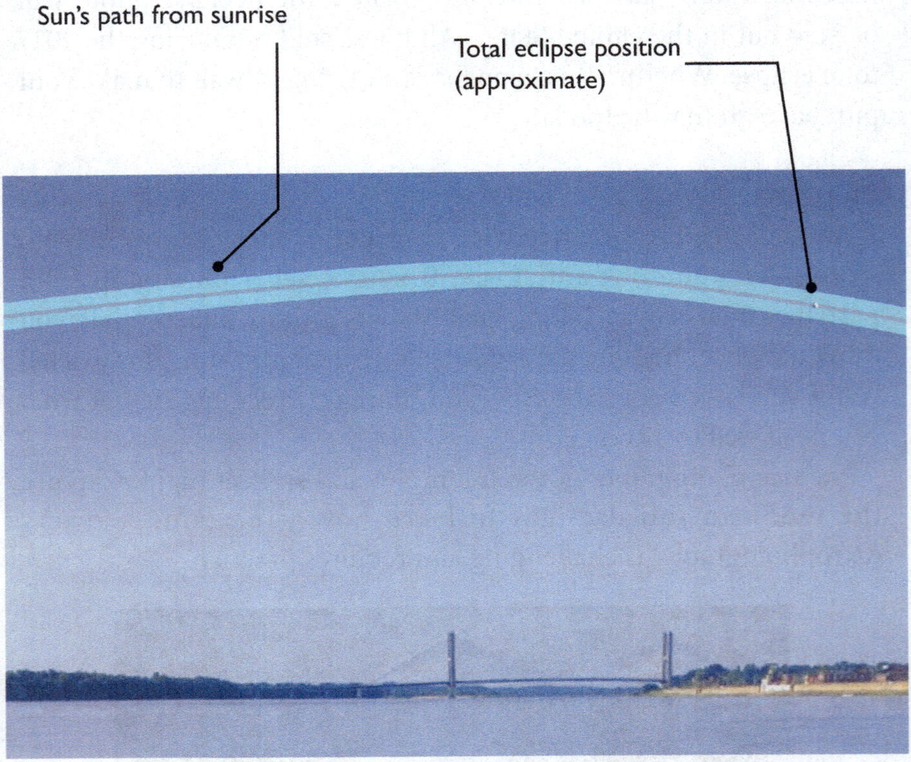

The eclipse will follow this approximate path on the afternoon of April 8, 2024.

Image of the Cape Girardeau Bill Emerson Memorial Bridge.

Note that this image is a simulation and approximation the sun's path and where the total eclipse may appear from one perspective. Refer to the eclipse position data for a more accurate location.

⊙ is the symbol for the sun and first appeared in Europe during the Renaissance.

☾ is the ancient symbol for the moon.

Viewing Locations Around Missouri

Tens of thousands of people will travel to and around Missouri to view the total eclipse. There are few obstructions and there is a vast amount of space to view the total eclipse from.

If the weather is questionable, seek out a new location as soon as possible. If you wait until the hour before the eclipse, you may find yourself stuck in traffic, as others will be looking for a viewing location. Be safe on the roadways, as drivers may be distracted.

This section contains popular, alternative, and little-known locations to watch the eclipse. As long as there are no clouds or smoke from fires, the partial eclipse will be viewable from anywhere in the state.

Suggested Total Eclipse View Points

Towns and Cities

- Cape Girardeau
- Dexter
- Doniphan
- Ellington
- Eminence
- Farmington
- Fredericktown
- Kennett
- Malden
- Marble Hill
- New Madrid
- Perryville
- Poplar Bluff
- Sikeston
- Ste. Genevieve
- West Plains
- Winona

Missouri Total Eclipse Path

Unique Locations

- Grand Gulf State Park
- Lake Wappapello
- Mark Twain NF
- Ozark NSR

Cape Girardeau

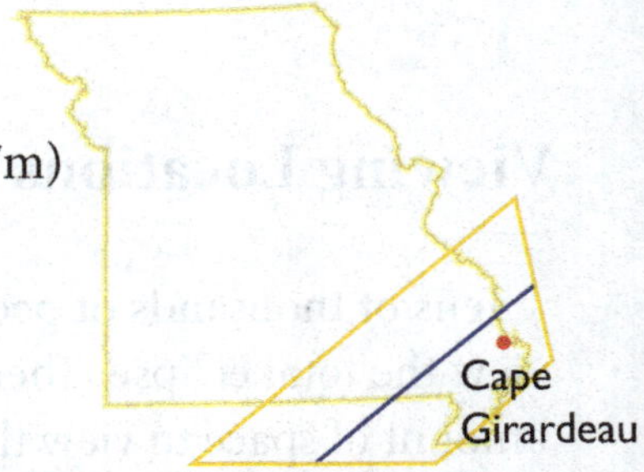

Elevation:	351 ft. (107m)
Population:	39,853
Main road/hwy:	I-55

Overview

Named after the French soldier Jean Baptiste de Girardot, Cape Girardeau is located 175 miles from Memphis and is the largest city in southeast Missouri. Apart from being known for its historical sites, murals, and cemeteries, the city is the home of Southeast Missouri State University and has earned the reputation of being an upcoming college town. Some of the most popular places in the city include the Mississippi River Tales Mural, Fort D, which is a Civil War-era fort alongside the Mississippi River, and the Confederate War Memorial.

Getting There

Drive south from St. Louis on I-55 for one hundred seventeen miles to reach Cape Girardeau.

Totality Duration

4 minutes 5 seconds

Notes

The city of Cape Girardeau's website is a good starting place to search for local total eclipse information. Visit it at https://www.cityofcapegirardeau.org/.

Event	Time (CDT)	Altitude	Azimuth
Sunrise	6:32:00AM	0°	79°
Eclipse Start	12:41:52PM	60°	171°
Totality Start	1:58:19PM	57°	207°
Totality End	2:02:25PM	57°	209°
Eclipse End	3:17:27PM	46°	235°
Sunset	7:27:00PM	0°	280°

Dexter

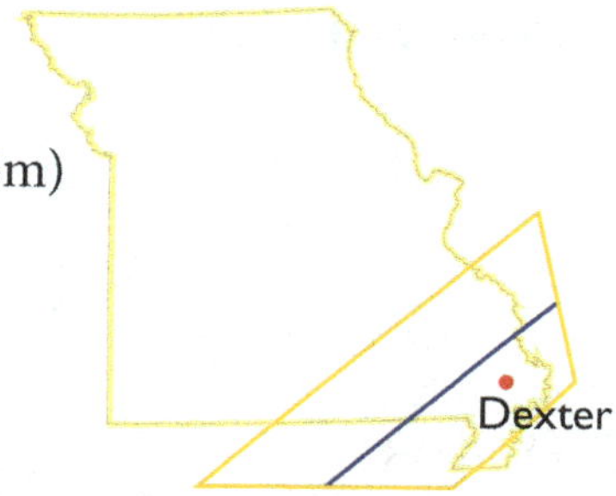

Elevation:	371 ft. (113m)
Population:	4,106
Main road/hwy:	US 60

Overview

Dexter, located in Stoddard Country, was founded in 1873. With numerous museums, it has had a rich history. This is evident from the two museums within the city limits as well as several others in its neighboring areas surrounding it. With its humid summer climate, Dexter is also popular for its seven wildlife conservation parks. These are a perfect getaway for people who love nature outdoors. You cannot run out of activities to do in and around Dexter. With a forest area of thirty-five thousand acres, it is the perfect place for travelers to go camping, hiking, hunting, fishing, and wildlife viewing.

Getting There

Drive south from St. Louis on I-55 for one hundred forty-two miles, then turn west on US 60 and drive twenty-five miles to reach Dexter.

Totality Duration

3 minutes 51 seconds

Notes

Visit the news website http://visitdexter.com/ for updates on events, lodging, and travel in Dexter.

Event	Time (CDT)	Altitude	Azimuth
Sunrise	6:34:00AM	0°	79°
Eclipse Start	12:40:24PM	60°	169°
Totality Start	1:57:10PM	58°	206°
Totality End	2:01:01PM	57°	208°
Eclipse End	3:16:29PM	47°	234°
Sunset	7:28:00PM	0°	281°

Doniphan

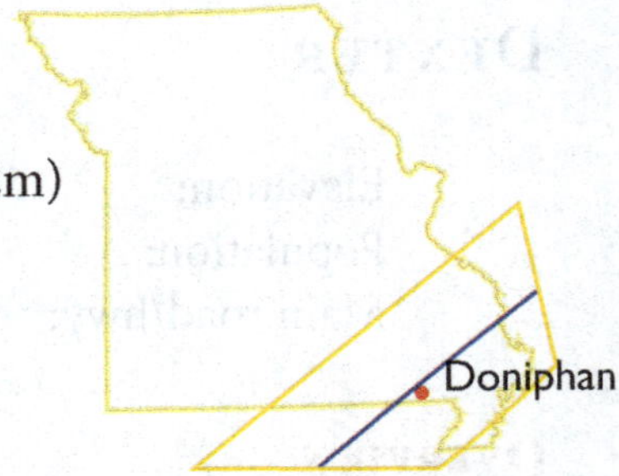

Elevation:	400 ft. (122m)
Population:	1,933
Main road/hwy:	US 67

Overview

Located thirty miles west of Poplar Bluff, Doniphan is a hill city in Ripley County, Missouri, which lies above the beautiful Current River. The small city was named after the nineteenth-century American attorney, soldier, and politician Alexander William Doniphan. It is located near the center of Ripley County on the east bank of the river. It soon became the county seat in 1847, making it the focal point of the area. With modern amenities and the small-town feel, it ensures a perfect balance for its residents and eclipse chasers.

Getting There

Doniphan can be reached by driving south from St. Louis for forty miles on I-55, then continuing on US 67 for one hundred thirty seven miles.

Totality Duration

4 minutes 12 seconds

Notes

The Ripley County Chamber website will be a good starting point for your eclipse adventure at https://www.ripleycountymissouri.org/.

Event	Time (CDT)	Altitude	Azimuth
Sunrise	6:38:00AM	0°	80°
Eclipse Start	12:38:48PM	60°	166°
Totality Start	1:55:25PM	58°	204°
Totality End	1:59:37PM	58°	206°
Eclipse End	3:15:10PM	48°	233°
Sunset	7:32:00PM	0°	280°

Ellington

Elevation:	663 ft. (202m)
Population:	962
Main road/hwy:	MO 21

Overview

Ellington, which Clint Huff Elling laid out in 1856, is a city in Reynolds County. The city is near Eminence, Van Buren, and Lester Ville. This makes it a popular stop for tourists who are traveling to the Ozark National Scenic Riverways. The city is also known for the Civil War Fortification at Barnesville. This fort was constructed in 1863 during the American Civil War. Its primary feature is a horseshoe-shaped redoubt. Reynolds County Museum is another popular destination for visitors who will be viewing the total eclipse.

Getting There

Drive south from St. Louis on I-55 and continue south on US 67 for sixty-three miles. Continue on MO 32, MO 21, MO 49/72 for another fifty-nine miles to reach Ellington.

Totality Duration

3 minutes 15 seconds

Notes

Learn more about Ellington at http://www.missourihistorictowns.com/missouri-historic-places/ellington-missouri-history-landmarks-barnesville-logans-creek.html.

Event	Time (CDT)	Altitude	Azimuth
Sunrise	6:38:00AM	0°	79°
Eclipse Start	12:39:35PM	59°	167°
Totality Start	1:56:24PM	58°	204°
Totality End	1:59:40PM	57°	205°
Eclipse End	3:15:24PM	48°	233°
Sunset	7:33:00PM	0°	280°

Eminence

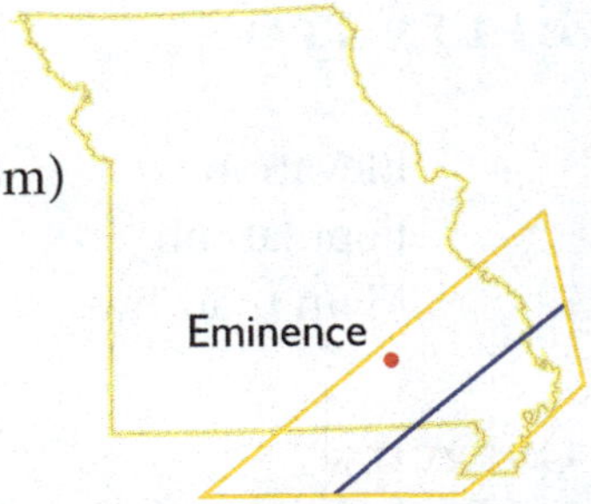

Elevation:	673 ft. (205m)
Population:	584
Main road/hwy:	MO 106

Overview

Founded over one hundred fifty years ago, Eminence is a small town on the banks of river Jacks Fork, known for its endless nature trails and love for country music. Located in the heart of Ozark National Scenic Riverways, it happens to be America's first national park created with the purpose of protecting a river system. It is also a very popular destination among horse lovers who enjoy spending time riding over the countryside. People also enjoy playing sports in the area, with fishing, golf, and tennis being particularly popular. If you are looking to experience an 1890s-style theme park, then you can also head to Silver Dollar City, an entertaining and educational place to hang out.

Getting There

Follow the directions from St. Louis to Ellington, then continue west on MO 106 for twenty-seven miles to reach Eminence.

Totality Duration

2 minutes 38 seconds

Notes

Visit https://www.visitmo.com/ for more information.

Event	Time (CDT)	Altitude	Azimuth
Sunrise	6:40:00AM	0°	80°
Eclipse Start	12:38:52PM	59°	166°
Totality Start	1:56:00PM	58°	203°
Totality End	1:58:39PM	58°	204°
Eclipse End	3:14:47PM	48°	232°
Sunset	7:34:00PM	0°	280°

Farmington

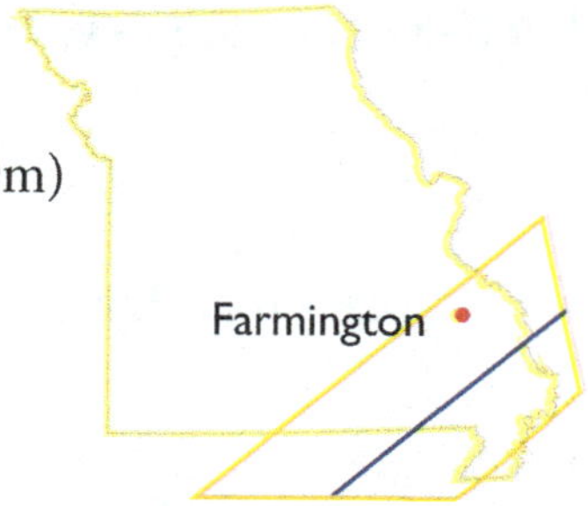

Elevation:	915 ft. (279m)
Population:	45,450
Main road/hwy:	US 67

Overview

Drive south of Saint Louis for an hour to discover the city of Farmington at the edge of scenic Ozark Mountains. It is surrounded by some of the most picturesque state parks in Missouri. The Parkland also features some of the state's finest locations to camp, fish, and hike. With a recreation area that stretches off the road, hiking trails, and fifteen award-winning wineries and vineyards, Farmington, hands down, is an ideal destination for anyone looking to spend time outdoors. If you are a golf lover, you'll enjoy the challenging golf courses at Saint Joe State Park.

Getting There

Drive south from St. Louis for seventy miles on I-55 and US 67 to reach Farmington.

Totality Duration

2 minutes 29 seconds

Notes

Point your web browser to https://discoverfarmingtonmo.com/ to plan your eclipse visit in Farmington.

Event	Time (CDT)	Altitude	Azimuth
Sunrise	6:35:00AM	0°	79°
Eclipse Start	12:41:16PM	59°	169°
Totality Start	1:58:15PM	57°	205°
Totality End	2:00:45PM	57°	206°
Eclipse End	3:16:31PM	47°	233°
Sunset	7:31:00PM	0°	280°

Fredericktown

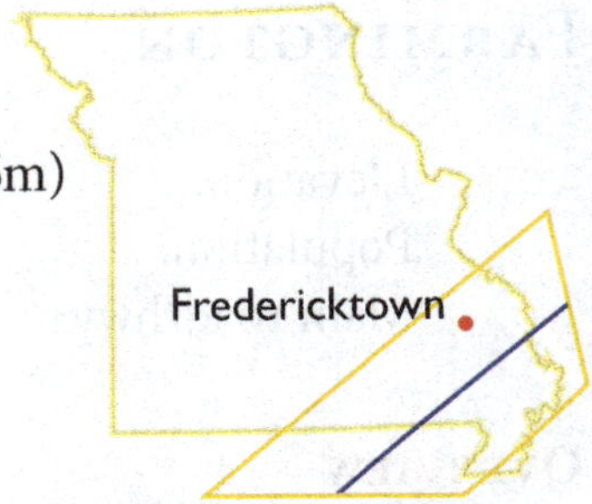

Elevation: 741 ft. (226m)
Population: 4,089
Main road/hwy: US 67

Overview

Surrounded by the Mark Twain National Forest at the foothills on the northeast side of splendid St. Francois Mountains is the city of Fredericktown, the county seat of Madison County. There are a lot of places in and around Fredericktown that are worth visiting. Many are worth seeing simply for the architecture, including the Fredericktown Courthouse Square District, Madison County Courthouse, Fredericktown Missouri Pacific Railroad Depot, and the Fredericktown United States Post Office.

Getting There

Drive south from St. Louis for eighty-eight miles on I-55 and US 67 to reach Fredericktown.

Totality Duration

3 minute 33 seconds

Notes

The Fredericktown website will have more information about eclipse-related events at http://fredericktownmo.org/.

Event	Time (CDT)	Altitude	Azimuth
Sunrise	6:35:00AM	0°	80°
Eclipse Start	12:41:06PM	59°	169°
Totality Start	1:57:40PM	57°	206°
Totality End	2:01:14PM	57°	207°
Eclipse End	3:16:33PM	47°	233°
Sunset	7:30:00PM	0°	280°

Kennett

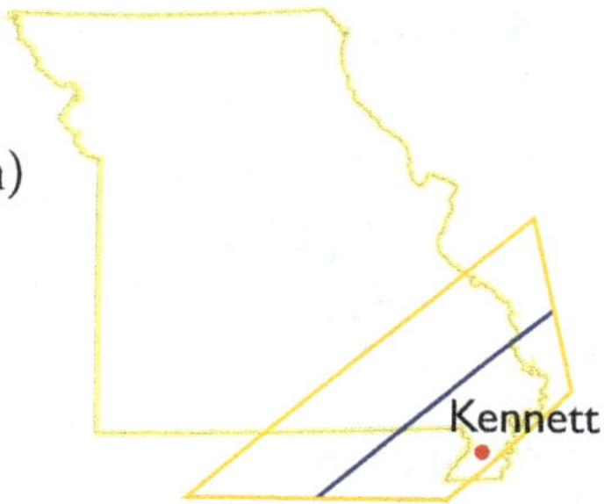

Elevation:	269 ft. (82m)
Population:	10,399
Main road/hwy:	US 412

Overview

Located twenty miles from the Mississippi River and adjoining America's most beautiful heartland is Kennett. This city is also the county seat of Dunklin County, making it a center of commerce for the area. It is primarily an agricultural area and boasts of being the largest producer of cotton, rice, watermelon, and cantaloupe in the state. Along with Grammy Award-winning artist Sheryl Crow, the city has also produced several country artists from David Nail to Trent Tomlinson. There is so much to do in Kennett, from food to shopping to the movies. If you are more adventurous, step outside the city limits to go hunting or fishing, a major tourist activity for those visiting.

Getting There

Drive south from St. Louis for one hundred ninety miles, then turn west on US 412 and continue for seventeen miles to reach Kennett.

Totality Duration

2 minutes 11 seconds

Notes

Visit Kennett's official website at http://www.visitkennett.com/.

Event	Time (CDT)	Altitude	Azimuth
Sunrise	6:35:00AM	0°	80°
Eclipse Start	12:39:22PM	61°	168°
Totality Start	1:57:10PM	58°	207°
Totality End	1:59:22PM	58°	208°
Eclipse End	3:15:56PM	48°	235°
Sunset	7:28:00PM	0°	280°

Malden

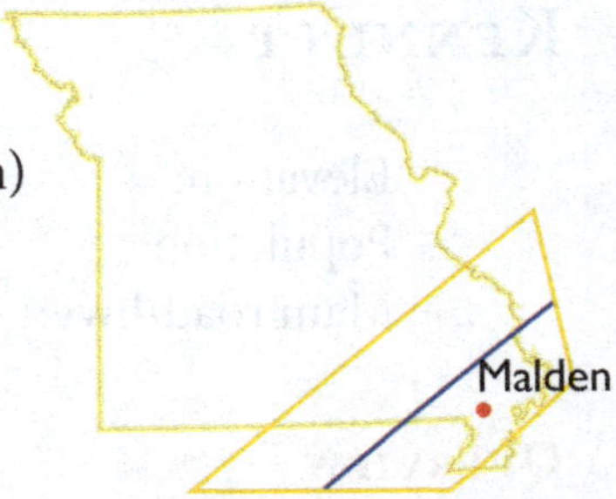

Elevation:	292 ft. (89m)
Population:	4,064
Main road/hwy:	US 62

Overview

Malden is a city in the northeast corner of Dunklin County, Missouri, and houses the Bootheel Museum, a unique activity museum. The Museum is one of the premium youth attractions in America and has won the National Award in the year 2016. With over one hundred forty thousand visitors, it is a resource for the school district. The Malden Historical Museum, the Malden Veterans Wall of Honor, and the Malden Community Center are some of the other popular attractions in the city.

Getting There

Drive north from Kennett for twenty-five miles on MO 25 to reach Malden.

Totality Duration

3 minutes 50 seconds

Notes

The Malden Army Airfield Preservation museum will have local event information. Visit them at https://www.visitmo.com/things-to-do/malden-maaps-museum or http://www.maaps.net/ for more information.

Event	Time (CDT)	Altitude	Azimuth
Sunrise	6:35:00AM	0°	79°
Eclipse Start	12:40:02PM	60°	169°
Totality Start	1:57:09PM	58°	207°
Totality End	2:00:28PM	58°	208°
Eclipse End	3:16:29PM	47°	235°
Sunset	7:28:00PM	0°	280°

Marble Hill

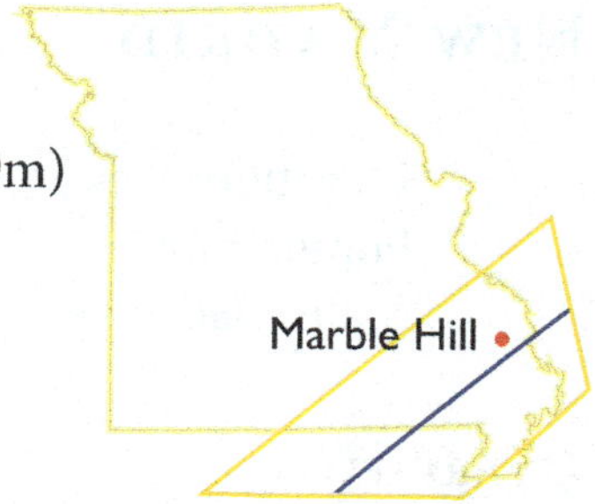

Elevation:	489 ft. (149m)
Population:	1,484
Main road/hwy:	MO 34

Overview

Marble Hill is the largest city in Bollinger County. The city was officially incorporated in July 1851 and takes its name from the nearby marble-like rock. If you are nearby this small town before or after the eclipse, take in the Bollinger County Veterans Memorial Wall and the Bollinger County Museum of Natural History for an educational excursion.

Getting There

Drive west from Cape Girardeau on I-55 Business Loop and MO 34 for thirty miles to reach Marble Hill.

Totality Duration

4 minutes 9 seconds

Notes

Start your exploration of Marble Hill at the Bollinger County SEMO GIS website at: https://semogis.com/bollingercounty.html.

Event	Time (CDT)	Altitude	Azimuth
Sunrise	6:34:00AM	0°	80°
Eclipse Start	12:41:11PM	59°	169°
Totality Start	1:57:35PM	57°	206°
Totality End	2:01:44PM	57°	208°
Eclipse End	3:16:50PM	47°	234°
Sunset	7:29:00PM	0°	280°

New Madrid

Elevation: 295 ft. (90m)
Population: 2,957
Main road/hwy: I-55 BL

Overview

New Madrid, founded in 1778 by a governor from Spain who was appointed to rule the land west of the Mississippi River, was a Spanish holding at the time. Located forty-two miles southwest of Cairo, Illinois, it boasts several attractions for its visitors. From the New Madrid Historical Museum to the scenic Mississippi River view, it provides plenty for a traveler to absorb and soak in. The town is also famous for the powerful 1811-1812 New Madrid earthquakes.

Getting There

Drive south from Cape Girardeau on I-55 for fifty-three miles and take the I-55 Business Loop to New Madrid.

Totality Duration

2 minutes 14 seconds

Notes

Visit New Madrid's website http://www.new-madrid.mo.us/ for updated eclipse event and lodging information.

Event	Time (CDT)	Altitude	Azimuth
Sunrise	6:33:00AM	0°	80°
Eclipse Start	12:40:44PM	60°	170°
Totality Start	1:58:24PM	58°	208°
Totality End	2:00:38PM	57°	209°
Eclipse End	3:16:55PM	47°	235°
Sunset	19:26:00PM	0°	280°

Perryville

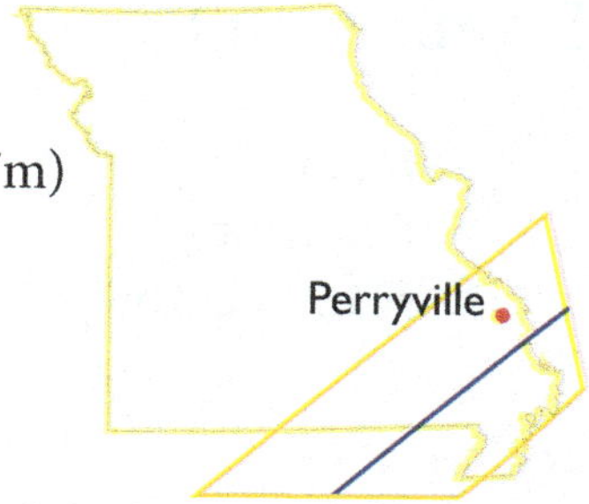

Elevation:	581 ft. (177m)
Population:	8,456
Main road/hwy:	US 61

Overview

The city of Perryville, which is often called the heartland's best-kept secret, is a city in Perry County that brings with it a rich confluence of culture from the French and Spanish administration of the eighteenth century. While the residents have managed to keep the essence of a small town intact, they have also managed to offer improved services and amenities to community residents as well as visitors. The historic locations provide a rich cultural backdrop to a city also known for its rural scenic landscape.

Getting There

Drive north from Cape Girardeau on I-55 for twenty-four miles, take exit 123, and continue north on Hwy B for seven miles to reach Perryville.

Totality Duration

3 minutes 46 seconds

Notes

Visit the Perry County website at https://www.perrycountymo.us/517/Tourism to begin your eclipse planning in Perryville.

Event	Time (CDT)	Altitude	Azimuth
Sunrise	6:33:00AM	0°	80°
Eclipse Start	12:42:00PM	59°	170°
Totality Start	1:58:25PM	57°	206°
Totality End	2:02:11PM	56°	208°
Eclipse End	3:17:15PM	46°	234°
Sunset	7:29:00PM	0°	280°

Poplar Bluff

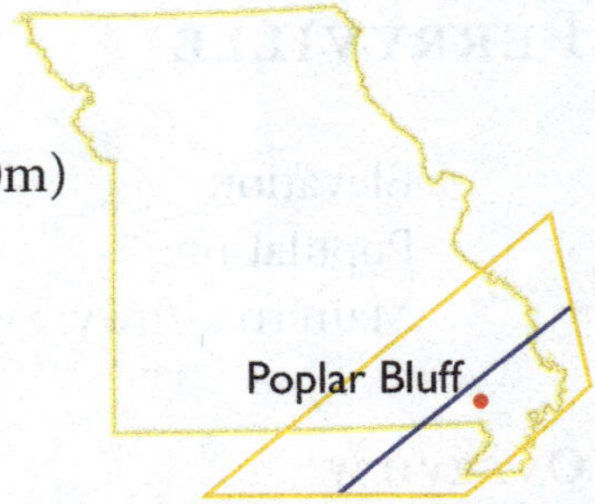

Elevation:	316 ft. (110m)
Population:	17,070
Main road/hwy:	US 60

Overview

A small city in Butler County in Southeast Missouri, Poplar Bluff is more popularly known as the "Gateway to the Ozarks." With tons of outdoor activities and natural beauty, the city is tucked between Little Rock, Memphis, and St. Louis, making it an amazing weekend getaway. It is popularly known for its award-winning barbeque and the unique pink lemonade. You could hike on one of several nature trails or simply hang out at the Clearwater Lake, which has several recreational activities on offer.

Getting There

Drive south from Cape Girardeau on I-55 and exit 66B to US 60 for a total of seventy-eight miles, then turn off on US 60 BUS W to reach Poplar Bluff.

Totality Duration

4 minutes 8 seconds

Notes

Start at the Butler County website for information on events in Poplar Bluff at https://visitbutlercountymo.com/.

Event	Time (CDT)	Altitude	Azimuth
Sunrise	6:36:00AM	0°	79°
Eclipse Start	12:39:41PM	60°	168°
Totality Start	1:56:18PM	58°	205°
Totality End	2:00:26PM	58°	207°
Eclipse End	3:15:52PM	48°	234°
Sunset	7:30:00PM	0°	280°

Sikeston

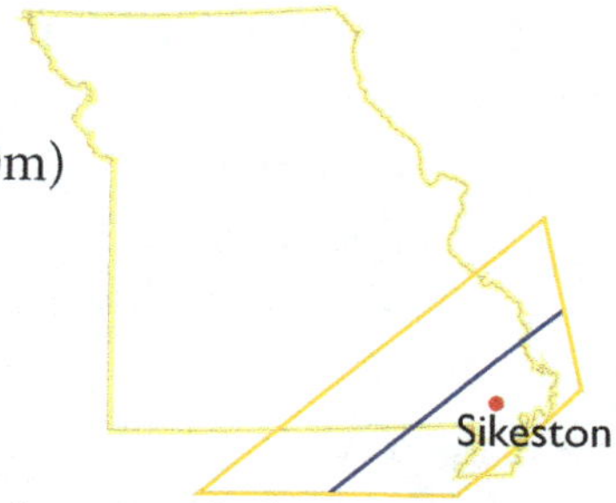

Elevation: 328 ft. (100m)
Population: 16,327
Main road/hwy: I-70

Overview

Three hours from Nashville, Tennessee, the city of Sikeston is located both in northern New Madrid County and southern Scott County. It is the fourth-most populous city in Missouri. It is particularly known for its park system, the largest and most popular of which is the Sikeston Recreation Complex and has a host of activities on offer. The complex also hosts a large number of sporting tournaments both at the state and regional level. You can also find cool picnic places, fishing spots, and playground equipment at the park to enjoy your time with the family while the total eclipse happens overhead.

Getting There

Drive south from Cape Girardeau on I-55 for thirty-one miles, then take US 60 Business loop to Sikeston.

Totality Duration

3 minutes 31 seconds

Notes

Sikeston's website is a great place to start your eclipse event planning. Visit it at https://www.sikeston.org/.

Event	Time (CDT)	Altitude	Azimuth
Sunrise	6:33:00AM	0°	80°
Eclipse Start	12:41:06PM	60°	170°
Totality Start	1:58:00PM	57°	207°
Totality End	2:01:31PM	57°	209°
Eclipse End	3:17:03PM	47°	235°
Sunset	7:27:00PM	0°	280°

Ste. Genevieve

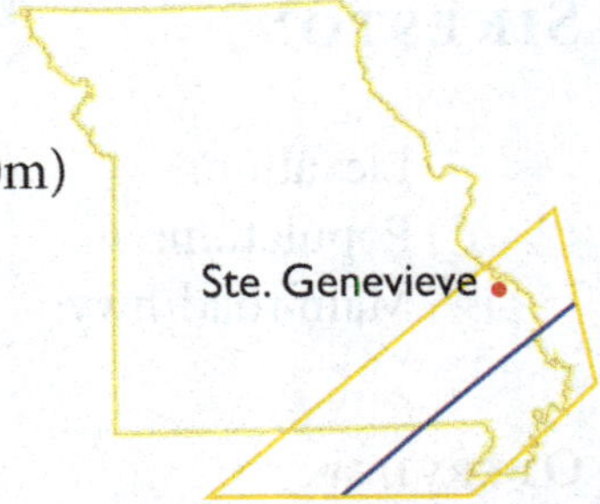

Elevation: 560 ft. (170m)
Population: 4,440
Main road/hwy: US 61

Overview

Ste. Genevieve was founded in 1735 by French Canadian colonists and settlers. It was named after Saint Genevieve, who lived in the fifth century A.D. (ACE) and was the patron saint of Paris, the capital of France. It's the first organized European settlement west of the Mississippi River in present-day Missouri and is filled with historic attractions. The eighteenth-century architecture provides glimpses into its colonial past, and you can feel the ambiance of the place as you walk through the streets of the city.

Getting There

Drive south from St. Louis on I-55 for fifty-three miles, then take US 61 east for eight miles to reach Ste. Genevieve.

Totality Duration

2 minute 40 seconds

Notes

Learn more about the city and total eclipse events at https://www.visitstegen.com/.

Event	Time (CDT)	Altitude	Azimuth
Sunrise	6:34:00AM	0°	79°
Eclipse Start	12:42:07PM	59°	170°
Totality Start	1:58:58PM	57°	206°
Totality End	2:01:38PM	56°	207°
Eclipse End	3:17:09PM	46°	233°
Sunset	7:30:00PM	0°	280°

LOCATIONS

West Plains

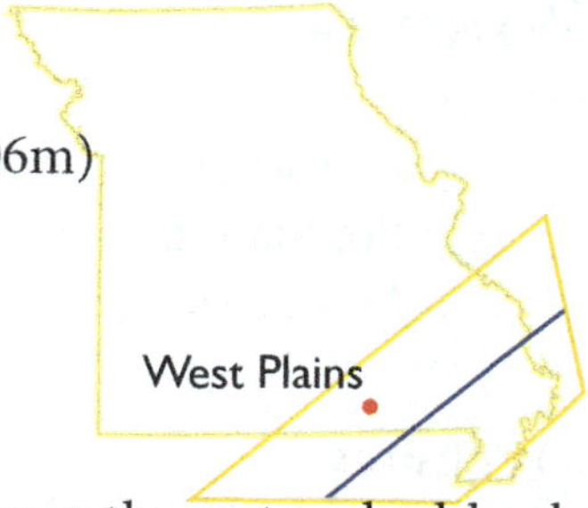

Elevation:	1,004 ft. (306m)
Population:	12,248
Main road/hwy:	US 63

Overview

Tucked within the Ozark Mountains near the untouched backwoods and spring-fed streams, the city of West Plains is a perfect retreat for nature lovers as well as families who are looking for big-city conveniences with small-town coziness. Even though it is the largest city within a hundred-mile radius of south-central Missouri, the city has its own pace and is the perfect place to experience life on your own terms.

Getting There

Drive east from Springfield, MO on US 60/63 for one hundred five miles to reach West Plains.

Totality Duration

3 minutes 2 seconds

Notes

Stop by the West Plains website to start your total eclipse planning at https://westplains.net/.

Event	Time (CDT)	Altitude	Azimuth
Sunrise	6:42:00AM	0°	80°
Eclipse Start	12:37:27PM	60°	164°
Totality Start	1:54:31PM	59°	202°
Totality End	1:57:33PM	58°	203°
Eclipse End	3:13:47PM	49°	232°
Sunset	7:36:00PM	0°	280°

WINONA

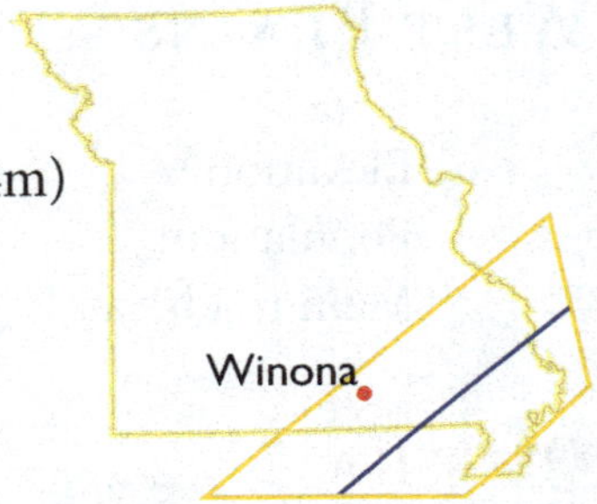

Elevation:	932 ft. (284m)
Population:	1,296
Main road/hwy:	US 60

OVERVIEW

Winona is a city in Shannon County. The population was 1,335 at the 2010 census and this small town is perfect for a farm-oriented vacation. If you want to escape crowds and busy streets to enjoy a quiet solo trip or some quality time on a farm, this small town is just right to enjoy America's last total eclipse for a generation.

GETTING THERE

Drive east from Springfield, MO on US 60 for one hundred thirty miles to reach Winona.

TOTALITY DURATION

3 minutes 16 seconds

NOTES

The Shannon County website is a good starting point for your total eclipse exploration in Winona at http://shannon-county.com/.

Event	Time (CDT)	Altitude	Azimuth
Sunrise	6:40:00AM	0°	80°
Eclipse Start	12:38:41PM	59°	165°
Totality Start	1:55:34PM	58°	203°
Totality End	1:58:50PM	58°	205°
Eclipse End	3:14:44PM	48°	232°
Sunset	7:34:00PM	0°	280°

Grand Gulf State Park

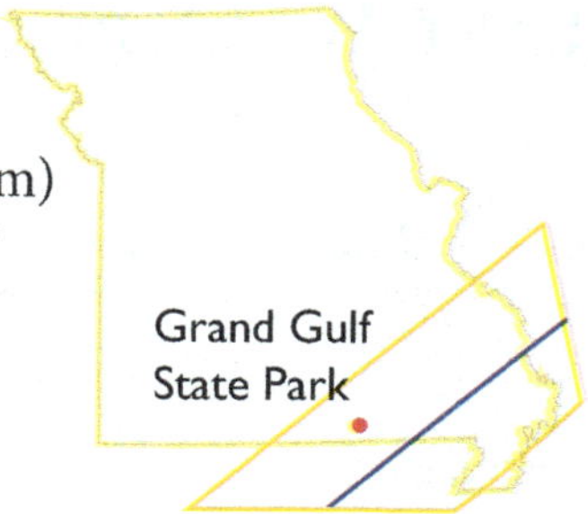

Elevation:	712 ft. (217m)
Main road/hwy:	Co. Rd. 377

Overview

Known as Missouri's Little Grand Canyon, this stunning park is a gem of the Ozarks. It features a natural stone bridge that has a 75-foot tall opening and is 250 wide. For those who are adventurous and don't mind some risk, you can bushwhack your way to the bottom. This park is a natural wonder, as it is a fascinating example of a collapsed cave system. This three-quarters of a mile long vertical wall canyon will provide unique total eclipse photography options.

Getting There

Drive southeast from West Plains on US 63 for fourteen miles, then turn south on County Road 377 and drive for four miles to reach the park.

Totality Duration

3 minutes 51 seconds

Notes

The park's website will have updated eclipse event information at https://mostateparks.com/park/grand-gulf-state-park.

Event	Time (CDT)	Altitude	Azimuth
Sunrise	6:41:00AM	0°	80°
Eclipse Start	12:37:27PM	60°	164°
Totality Start	1:54:13PM	59°	202°
Totality End	1:58:04PM	58°	204°
Eclipse End	3:13:56PM	49°	232°
Sunset	7:35:00PM	0°	280°

Lake Wappapello

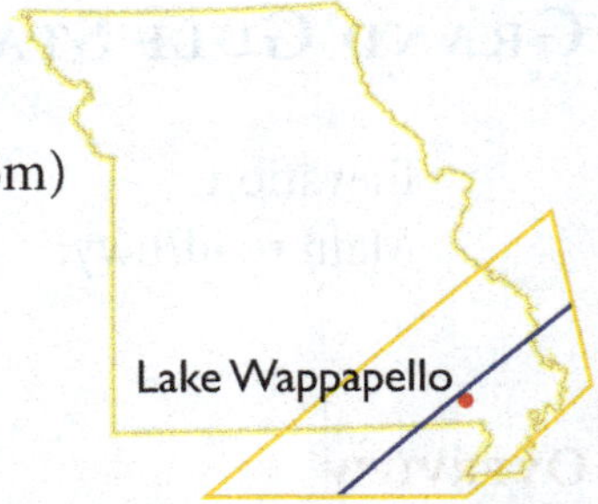

Elevation: 377 ft. (115m)
Main road/hwy: MO 172

Overview

Located one hundred twenty miles south of St. Louis, Lake Wappapello is a reservoir on the St. Francis River, featuring a park and recreational area accessible to the public with boating, fishing, and swimming. Camp overnight at the State Park in one of the many rustic cabins, and wake up to nature at its best.

Getting There

Drive northwest from Poplar Bluff on US 60/67 for sixteen miles, then turn east onto MO 172 and continue four miles to reach Lake Wappapello.

Totality Duration

4 minutes 10 seconds

Notes

Visit Lake Wappapello's website for updated eclipse event information at https://mostateparks.com/park/lake-wappapello-state-park.

Event	Time (CDT)	Altitude	Azimuth
Sunrise	6:35:00AM	0°	80°
Eclipse Start	12:40:06PM	60°	168°
Totality Start	1:56:37PM	58°	205°
Totality End	2:00:48PM	57°	207°
Eclipse End	3:16:07PM	47°	234°
Sunset	7:30:00PM	0°	280°

Mark Twain National Forest Ozark

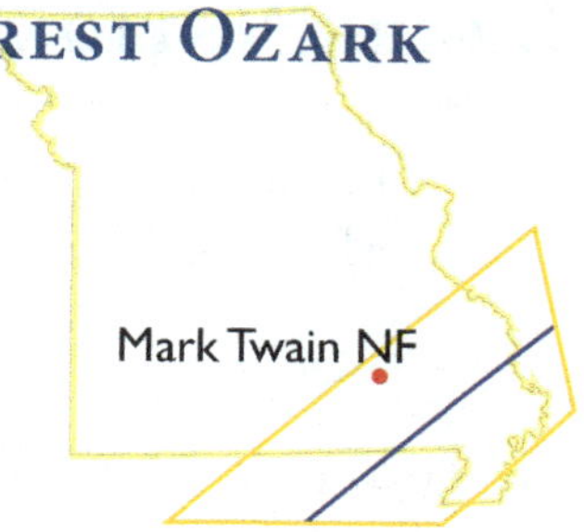

Elevation:	Varies
Main road/hwy:	Varies

Overview

Named after the famous writer Mark Twain, a native of Missouri, the Mark Twain National Forest is an area located in the southern half of Missouri. Plan your day of biking, horseback riding, swimming or hiking the famous three hundred fifty-mile Ozark Trail that winds through the national forest. The park is home to diverse flora and fauna and a wide variety of animals. If you are a nature lover, this is the place to be for the total eclipse.

Getting There

See the forest website for directions: www.fs.usda.gov/main/mtnf

Totality Duration

3 minutes 29 seconds

Notes

The Mark Twain National Forest is Missouri's only national forest. There are other woodlands administered by other government agencies.

Chart times are for Womack, MO. Other location times vary.

Event	Time (CDT)	Altitude	Azimuth
Sunrise	6:34:00AM	0°	79°
Eclipse Start	12:41:26PM	59°	169°
Totality Start	1:58:02PM	57°	206°
Totality End	2:01:26PM	57°	207°
Eclipse End	3:16:46PM	47°	207°
Sunset	7:30:00PM	0°	281°

Ozark National Scenic Riverways

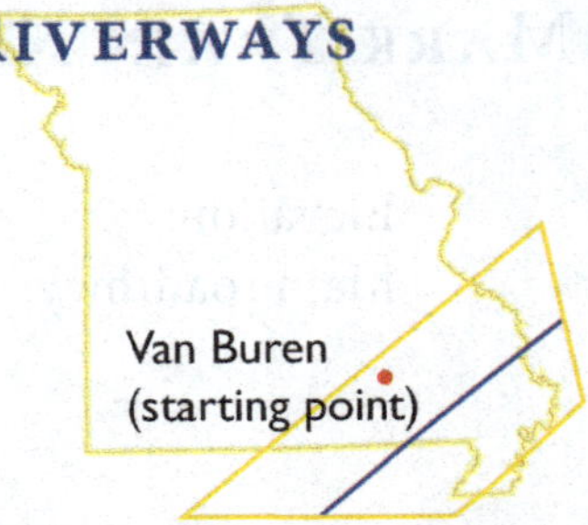

Elevation: Varies
Main road/hwy: Varies

Overview

Formed by an Act of Congress in 1964 with an aim to protect the Current and Jacks Fork Rivers, the Ozark National Scenic Riverways is a popular destination among nature lovers. When joined to the Arkansas Ozark Highlands Trail, the full end-to-end hiking distance is at least seven hundred miles. Spread over 80,000 acres, it hosts several recreational activities such as hiking, backpacking, biking, and equestrian trail. Canoeing is another popular activity, which is a big hit among tourists and travelers. The river will offer unique total eclipse viewing opportunities. Also, you can experience the event in solitude along the riverways in this national scenic area.

Getting There

Start your exploration in Van Buren, MO along the Current River.

Totality Duration

Varies depending on location.

Notes

Visit the National Park's website to learn more about the Ozark National Scenic Riverways at https://www.nps.gov/ozar/index.htm.

Remember the Missouri Total Eclipse

April 8, 2024

Who was I with?

What did I see?

What did I feel?

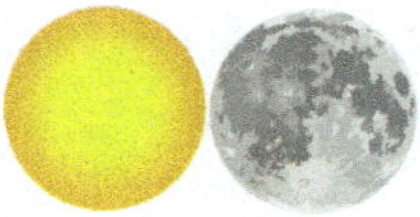

What did the people with me think?

Where did I stay?

Enjoy Other Books by Aaron Linsdau

50 Jackson Hole Photography Hotspots

This guide reveals the best Jackson Hole photography spots. Learn what locals and insiders know to find the most impressive and iconic photography locations in the United States. This is an excellent companion guide to the *Jackson Hole Hiking Guide*.

www.sastrugipress.com/books/50-jackson-hole-photography-hotspots/

Adventure Expedition One
by Aaron Linsdau M.S. & Terry Williams, M.D.

Create, finance, enjoy, and return safely from your first expedition. Learn the techniques explorers use to achieve their goals and have a good time doing it. Acquire the skills, find the equipment, learn to camp, understand medical issues, and learn the planning necessary to pull off an expedition.

www.sastrugipress.com/books/adventure-expedition-one/

Antarctic Tears

Experience the honest story of solo polar exploration. This inspirational true book will make readers both cheer and cry. Coughing up blood and fighting skin-freezing temperatures were only a few of the perils Aaron Linsdau faced. Travel with him on a world-record expedition to the South Pole.

www.sastrugipress.com/books/antarctic-tears/

How to Keep Your Feet Warm in the Cold

Keep your feet warm in cold conditions on chilly adventures with techniques described in this book. Packed with dozens and dozens of ideas, learn how to avoid having cold feet ever again in your outdoor pursuits.

www.sastrugipress.com/books/how-to-keep-your-feet-warm-in-the-cold/

Jackson Hole Hiking Guide

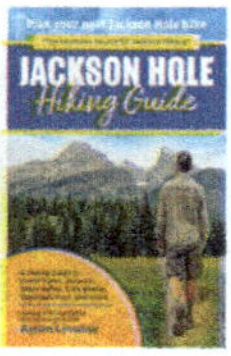

Jackson Hole contains some of the most dramatic and iconic landscapes in the United States. The book shares everything you need to know to hike Jackson's classic trails with canyons, high mountains, and hidden alpine lakes. This book is an excellent companion guide to *50 Jackson Hole Photography Hotspots*.

www.sastrugipress.com/books/jackson-hole-hiking-guide/

Subscribe to Aaron's YouTube channel at www.youtube.com/@alinsdau

If you enjoyed this book, please consider leaving a five-star review and a few words on what you liked about it at your favorite online retailer.

Lost at Windy Corner

Windy Corner on Denali has claimed fingers, toes, and even lives. What would make someone brave lethal weather, crevasses, and avalanches to attempt to summit North America's highest mountain? Aaron Linsdau shares the experience of climbing Denali alone and how you can apply the lessons to your life.
www.sastrugipress.com/books/lost-windy-corner/

The Motivated Amateur's Guide to Winter Camping

Winter camping is one of the most satisfying ways to experience the wilderness. It is also the most challenging style of overnighting in the outdoors. Learn 100+ tips from a professional polar explorer on how to winter camp safely and be comfortable in the cold.
www.sastrugipress.com/books/the-motivated-amateurs-guide-to-winter-camping/

Two Friends and a Polar Bear
by Terry Williams, M.D. & Aaron Linsdau

This story of friendship is about two old friends who plan to ski across the Greenland Ice Cap along the Arctic Circle in hopes of becoming one of the oldest teams to succeed.
www.sastrugipress.com/books/two-friends-and-a-polar-bear/

About the Author

Aaron Linsdau is the second American to ski alone from the coast of Antarctica to the South Pole (730 miles / 1174 km), setting a world record for surviving the longest expedition ever for that trip. He lead a 310-mile (499 km) ski expedition across the Greenland icecap along the Arctic Circle. Aaron has climbed Denali solo, crossed the Greenland tundra alone, skied across Yellowstone National Park solo, trekked through the Sahara desert, and successfully climbed Mt. Kilimanjaro and Mt. Elbrus in Russia.

Aaron Linsdau at the South Pole.

Use your smart device to scan the QR codes for website links.

Visit www.aaronlinsdau.com/subscribe to learn more about the author. Receive updates when he releases new books and shows.

Visit Sastrugi Press on the web at www.sastrugipress.com to purchase the above titles in bulk. They are available in print, e-book, or audiobook form.

Thank you for choosing Sastrugi Press.

Enjoy Other Books by Sastrugi Press

50 Florida Wildlife Hotspots by Moose Henderson Ph.D.

This is a definitive guide to finding where to photograph wildlife in Florida. Follow the guidance of a professional wildlife photographer as he takes you to some of the best places to see wildlife in the Sunshine State.

www.sastrugipress.com/books/50-florida-wildlife-hotspots/

50 Wildlife Hotspots Grand Teton National Park by Moose Henderson Ph.D.

Find out where to find animals and photograph them in Grand Teton National Park from a professional wildlife photographer. Learn techniques, timing, animal behavior, and composition to create stunning wildlife images.

www.sastrugipress.com/books/50-wildlife-hotspots/

Alaska: A Guide for the Curious by Nikki Mann & Jeff Wohl

Discover the natural world of Alaska. Find out what the plants and animals are like, how to identify them, and what the environment of Alaska is like.

www.sastrugipress.com/books/alaska-a-guide-for-the-curious/

Blood Justice by Tim W. James

Two brothers, one a preacher's son, the other an adopted would-be slave, set out in opposite directions to avenge their family's murder only to cross paths in pursuit of the killer. Book 1 of the Roger Brinkman Series.

www.sastrugipress.com/iron-spike-press/blood-justice/

Shake Yourself Free by Bob Millsap

Learn how to overcome difficult encounters with misfortune, tragedy, and loss. Emotional recovery is a journey requiring a mindset shift. Get this book now and take control of your life.

www.sastrugipress.com/books/shake-yourself-free/

The Burqa Cave by Dean Petersen

Still haunted by Iraq, a retired soldier seeks solace teaching high school in Wyoming. He soon finds the quiet town is home to murderers, maniacs, and a boy who can see where missing murder victims are. This paranormal thriller-romance surprises readers with unexpected twists.

www.sastrugipress.com/books/the-burqa-cave/

Photography Notes

April 8, 2024

Settings ______________________________

Successes ______________________________

Challenges ______________________________

Photography Notes

April 8, 2024

Camera

Company

Location

Made in the USA
Monee, IL
03 November 2026

67381727R00046